ELECTRO-CHEMISTRY

Inorganic

MAVEN BOOKS

ELECTRO-CHEMISTRY

Inorganic

G. GORE, F.R.S., LL.D.,

MAVEN BOOKS

Chennai   Trichy   Tirunelveli   New Delhi

MAVEN BOOKS

An Imprint of **MJP Publishers**

ISBN 978-93-87867-65-9 **MAVEN Books**

All rights reserved No. 44, Nallathambi Street,
Printed and bound in India Triplicane, Chennai 600 005

MJP 772 © Publishers, 2020

Publisher : **C. Janarthanan**

PUBLISHER'S NOTE

The legacy of a country is in its varied cultural heritage, historical literature, developments in the field of economy and science. The top nations in the world are competing in the field of science, economy and literature. This vast legacy has to be conserved and documented so that it can be bestowed to the future generation. The knowledge of this legacy is slowly getting perished in the present generation due to lack of documentation.

Keeping this in mind, the concern with retrospective acquiring of rare books has been accented recently by the burgeoning reprint industry. MAVEN Books is gratified to retrieve the rare collections with a view to bring back those books that were landmarks in their time.

In this effort, a series of rare books would be republished under the banner, "MAVEN Books". The books in the reprint series have been carefully selected for their contemporary usefulness as well as their historical importance within the intellectual. We reconstruct the book with slight enhancements made for better presentation, without affecting the contents of the original edition.

Most of the works selected for republishing covers a huge range of subjects, from history to anthropology. We believe this reprint edition will be a service to the numerous researchers and practitioners active in this fascinating field. We allow readers to experience the wonder of peering into a scholarly work of the highest order and seminal significance.

MAVEN Books

INDEX TO CONTENTS.

INTRODUCTION.

No separate treatise on Electro-Chemistry exists in the English language. The facts relating to the subject lie scattered in a great number of books and periodicals. Perceiving the utility of such a treatise, I have collected the numerous truths yet discovered in the subject and arranged them in consecutive order in the following pages. The treatise is not, however, merely a systematic and orderly collection of facts, but contains also brief descriptions of the known laws and general truths which underlie them.

The scope of the treatise is limited to the Electro-Chemistry of what is conventionally termed *mineral* compounds. Whilst nearly all the ordinary liquid and liquefiable salts belonging to "inorganic" chemistry have been subjected to the action of an electric current, and the effects observed, the influence of the current upon "organic" substances, although a subject of great extent, has hitherto been comparatively little investigated, and the facts as yet obtained in "organic" electrolysis are of an isolated and fragmentary character.

As the purpose for which the matter of this book was originally written rendered it advisable to limit the scope of the subject and to compress a large amount of information into a small compass, the laws and principles of the subject are only briefly illustrated.

The present treatise is essentially a Scientific one, and all facts and information of a purely Technical character have been purposely omitted.

G. GORE.

Birmingham, 1885.

ELECTRO-CHEMISTRY.

INORGANIC.

THE present series of articles is intended to contain, in systematic order, the chief principles and facts of electro-chemistry, and to supply to the student of electro-plating or electro-metallurgy a scientific basis upon which to build the additional practical knowledge and experience of his trade. As the series is a purely scientific one, it will not include such technical details or particulars as will enable the practical worker to obtain perfect workshop results ; these may be obtained from technical books on electro-metallurgy, combined with actual workshop experience. A scientific foundation, such as is here given, of the art of electro-metallurgy, is, however, indispensable to the electro-depositor who wishes to excel in his calling, and should be studied previously to and simultaneously with practical working. It is partly in consequence of deficiency of such fundamental knowledge by the British workman (and partly to the undue pursuit of wealth by his employers) that English manufactures are gradually being transferred to foreign lands. Whilst, also, the series of articles will contain the chief facts upon which the comparatively new art of electro-chemical analysis of minerals and alloys is based, it will not supply the technical details necessary for the accurate quantitative determination of metals by electro-chemical processes ; references to sources of such information will, however, be given.

The molecular weights of substances, as given at the heads of the paragraphs, are in nearly all cases those of the anhydrous ones ; for those of the hydrated compounds the student is referred to books on chemistry.

History.—The history of electro-chemistry requires only a brief description. Ages before the discovery of voltaic electricity it was known that various metals, by being simply immersed in metallic solutions, became coated with the metal previously dissolved in the liquid. Thousands of years ago Zosimus mentioned the deposition of bright metallic copper upon iron immersed in a solution of a salt of copper. In the year 1752 Sulzer remarked, "If you join two pieces of lead and silver, so that they shall be upon the same plane, and then lay them upon the tongue, you will notice a certain

B

taste resembling that of green vitriol, while each piece apart produces no such sensation." Paetz and Van Troostvik also, in the year 1790, decomposed water by passing electric sparks through it by means of very fine gold wires.

It was, however, the discovery by Volta, in 1799, of his electric battery which gave the first great impulse to electro-chemistry. By means of it Nicholson and Carlisle first decomposed water by means of a voltaic current from a battery on May 2nd, 1800 ; and soon afterwards Dr. Henry, of Manchester, decomposed nitric and sulphuric acids, and also ammonia by similar means. During the next year Dr. Wollaston discovered that if a piece of silver in connection with a more positive metal be put into a solution of copper the silver becomes coated with copper, which coating will stand the operation of burnishing. In the year 1801 Gerboin also first noticed the movements produced in mercury during the act of electrolysis. In 1803 Hisinger and Berzelius discovered that by means of a voltaic current the elements of water and of neutral salts were transferred to the respective polar wires immersed in the liquid ; and Cruickshank, about the same time, observed the electro-deposition of lead, copper, and silver upon one of the polar wires (the one connected with the zinc end of the battery) immersed in solutions of salts of those metals, and was thus led to suggest the analysis of minerals by means of the voltaic current. In 1805 Brugnatelli observed the electro-deposition of gold upon silver when the former was made the negative pole in a solution of ammoniuret of gold ; he also discovered the electro-deposition of zinc.

The most striking proof, however, of the great chemical power of the electric current was the discovery, on October the 6th, 1807, by Sir Humphrey Davy, of the electrolytic decomposition of potash and soda, and the liberation of their respective metals, by a current from a voltaic battery composed of 274 cells. In 1826 Nobili discovered the deposition of peroxide of lead in films of beautiful colour upon the platinum plate which conveyed a voltaic current into a solution of acetate of lead, and Ritter subsequently discovered the deposition of peroxide of silver from a solution of argentic nitrate under similar conditions. In 1834 Faraday discovered the important truth that, by the passage of an electric current through an undivided series of solutions of various metallic salts, or through those salts whilst in a state of fusion, the quantity of each salt decomposed was in direct proportion to the amount of current. Also that the quantities of the different metals dissolved or deposited were in definite proportions by weight, and that those proportions were identical with those of the ordinary chemical equivalents of those metals ; and he thus established the law of definite

electro-chemical action. And in 1837 Dr. Golding Bird succeeded in decomposing, by means of feeble voltaic currents,
solutions of the chlorides of sodium and potassium, and
depositing their respective metals into mercury.

As the subject of electro-chemistry is a very large one, it
is only briefly treated in the following series of articles. The
general principles and phenomena will be first explained, and
then will follow an account of the action of the current upon
individual substances.

Definition of Electro-Chemical Action.—Electro-chemical
action is chemical change produced by means of an electric
current, and usually consists of the decomposition of a compound liquid, the liquid being resolved into its constituent
parts in certain definite proportions by weight ; it is also
often attended by chemical union of a metal, in certain definite
proportions by weight, with one of the elements of the liquid.
It is usually limited to combinations of conducting substances
only.

Chief Conditions of Electro-Chemical Action.—The chief
conditions are that the substance must be a liquid, a compound
body, a conductor of electricity, and traversed by the current.
The liquids decomposable by a current are usually composed
of two elementary substances, the one being a metal and the
other a non-metal. Liquid alloys, or liquids composed of two
non-metallic elements, are not usually decomposed. Mixtures
of compounds in solution are commonly decomposed more
easily than solutions of single compounds; for instance, water
containing sulphuric acid is decomposed much more readily
than water alone.

All the products of electrolysis are set free in an almost
infinitely thin layer at the immediate surfaces of the conductors
at the parts where the current enters and leaves the liquid.
The electro-negative products, such as the non-metallic
elements and acids, are either liberated at or combine with
the conductor, by which the current enters the liquid, and
the electro-positive ones, such as metals and alkalies, are
liberated at or combine with the conductor by which the
current leaves the liquid. The behaviour of individual
combinations of metals and liquids will be subsequently
described.

Conductivity of Liquids.—Liquids present an extremely
wide range of conducting power ; whilst some completely
resist the passage of a current from 10,000 voltaic cells in
single series, others transmit freely the current from a single
element. Amongst the non-conducting ones may be included
all oils, benzine, petrolene, bisulphide of carbon, the liquid
chlorides of carbon, terchloride and pentachloride of phosphorus, terchloride of arsenic, pentachloride of antimony,

tetrachloride of tin, zinc-ethyl, perfectly pure water, bromine, various liquefied gases, including chlorine, carbonic anhydride, cyanogen, sulphurous anhydride, hydrochloric, hydrobromic, and hydriodic acids; nearly all melted fats and resins, fused iodine, sulphur, phosphorus, realgar, &c. Amongst the inferior conductors are aqueous solutions of gum, sugar, ammonia, boracic acid, mercuric cyanide, and alcoholic solutions of metallic salts; also melted boracic acid and fused glass. Amongst the best conducting compound liquids are aqueous solutions of salts of the alkali metals, and of copper, silver and gold, and especially certain fused salts, *e.g.*, argentic fluoride and chloride.

According to Hittorf, the degree of resistance of a liquid to electrolysis is dependent upon the difficulty with which the molecules exchange their constituents. Those which have active chemical properties should therefore conduct the best. Bleekrode contests this view.

Nomenclature.—The electrical decomposition of liquids is termed *electrolysis ;* the conductor by which the current is said to enter the liquid is called the *anode*, and the one by which it leaves it is termed the *cathode*. The products into which the liquid is decomposed are called *ions*, those which appear at the anode being *anions*, and those at the cathode *cations*. Non-metals, acids, and peroxides are usually anions, while metals and alkalies are cations; electro-negative bodies, therefore, usually appear at the anode or positive pole, and electro-positive ones at the cathode or negative pole. The same elementary substance, however, may appear at the positive pole in one case, and at the negative pole in another, according to the circumstance whether the body it is combined with is more positive or more negative than itself. For instance, iodine, when combined with a more positive body, such as hydrogen in hydriodic acid, appears at the anode; but when combined with a more negative one, such as oxygen in iodic acid, it appears partly at the cathode. Sulphur in suitable different combinations exhibits the same variation. Hydrogen is almost the only gaseous cation.

Visible Phenomena of Electrolysis. — The phenomena usually seen in a liquid during electrolysis are—at the anode, corrosion with or without solution of the anode, gas is evolved, the anode acquires an insoluble coating, &c. In some liquids the anode becomes fragile, and falls to powder; in others it flies to pieces, but this is a rare case. Silver in dilute hydrofluoric acid is an example of the former, and wood charcoal in anhydrous hydrofluoric acid is an instance of the latter. At the cathode, a soluble substance is set free and dissolves, or a gas, a liquid, or a solid is liberated, and is either absorbed by the cathode, or adheres to it, or is dissolved by the liquid, or escapes. The layers of liquid also in contact with the electrodes frequently alter in specific gravity, that at the anode usually becomes heavier, and descends, and that at the cathode lighter, and ascends.

Faraday, by passing an electric current upwards through a strong solution of "Epsom salt" into a layer of distilled water lying upon it, observed that a layer of magnesia formed at the upper surface of the lower liquid where it touched the water, as if the water acted as a cathode. Daniell also subsequently passed an upward current of electricity through solutions of the nitrates of silver, mercury, and lead, and of the sulphates of palladium, copper, iron, and magnesium, into a dilute one of caustic potash, separated from them by a thin horizontal diaphragm of bladder. Oxygen was determined to the upper, and the respective metals to the lower surface of the diaphragm, and coatings of metal, more or less oxidised, were formed against the latter surface, the oxidation being more complete the more oxidable the metal; with the magnesic solution a coating of oxide alone was formed. More recently (see *Proceedings of the Royal Society*, Nos. 212, p. 84, and 217, 1881, p. 142) I have stated, and shown by suitable experiments, that "every inequality of composition or of internal structure of the liquid in the path of the current must act to some extent as an electrode," and have also shown that a variety of phenomena take place at such a surface of mutual contact of two liquids when an electric current passes through it.

Movements of Liquid Electrodes.—As early as the year 1801 Gerboin observed the peculiar twitching movements of mercury whilst undergoing electrolysis, and which are now known to be due to electro-chemical action, and subsequently Sir H. Davy, Sir J. Herschel, and others investigated them. These movements are due to the formation and destruction, attended by contraction and expansion, of films upon the mercury, and would probably occur with other metallic electrodes whilst in the liquid state in suitable liquids. (*See* Gmelin's "Handbook of Chemistry," Vol. I., pp. 381-384.)

Sounds Emitted during Electrolysis.—Whilst investigating these peculiar movements and the thermic changes of electrolysis I discovered that in certain liquids a humming sound is emitted by electrodes of mercury, and that the surface of the mercury is covered with minute waves during the passage of the current. I also found that the current was intermittent during these vibrations (see *Proceedings of the Royal Society*, 1862). Sounds are not unfrequently emitted also from other metals whilst depositing, *e.g.*, from antimony. These are sometimes produced by contraction and cracking of the metals, at other times by explosion of bubbles of gas.

Decomposability of Electrolytes.—The degrees of facility with which different electrolytes are decomposed are different. Faraday has given the following order, the first-named being the easiest :—Solution of potassic iodide, fused chloride of silver, of zinc, of lead, melted iodide of lead, hydrochloric acid, dilute sulphuric acid. Smee gives nitric acid, solution of chloride of gold, nitrate of palladium, chloride of platinum, argentic nitrate, cupric sulphate, stannic sulphate, dilute sulphuric acid, solutions of the sulphates of cadmium, zinc, nickel, iron, and magnesium, and those of salts of the alkalies generally. Dilute sulphuric acid offers less resistance to electrolysis than one of zinc sulphate, and more than one of cupric sulphate (Favre, *Comptes Rendus*, Vol. LXXIII. ; *Journal of the Chemical Society*, 2nd series, Vol. X., p. 113). I have repeatedly observed that hydrochloric acid is decomposed more readily than water, and water more easily than hydrofluoric acid, also a solution of selenic acid before one of selenate of nickel. The readiness of decomposability of an electrolyte depends upon several conditions, and especially upon the nature of the electrodes ; thus a solution of potassic cyanide is readily decomposed when the anode is composed of palladium, silver, or copper, but with difficulty when it is formed of iron or platinum. A large field of research exists in this part of the subject.

The decomposability of a liquid is usually increased by rise of temperature ; it is also influenced by the length of the liquid portion of the circuit, which is the part in which the greatest resistance exists to the passage of the current. This has been shown by Gladstone and Tribe, who decomposed water by immersing in it a pure zinc plate previously coated electrolytically with a loose deposit of spongy copper or platinum, when two plates of those metals connected together and immersed at a distance from each other in the liquid would not decompose it, and have thus shown that "the dissociation of a binary compound may take place at infinitesimally short distances, when it would not take place where the layer of liquid is enough to offer resistance to the current" (*Proceedings of the Royal Society*, Vol. XX., p. 219).

According to Helmholtz, electrolysis of water by a voltaic current is possible only when the chemical processes in the battery, taken together, can produce more heat than the oxygen and hydrogen generated in the voltameter, and therefore that about $1\frac{3}{4}$ Daniell cells are required for a continuous decomposition of water. A single Daniell connected with platinum electrodes in dilute sulphuric acid produces only polarisation, no visible decomposition, the voltameter acting as a condenser of immense capacity (*Journal of the Chemical Society*, 2nd series, Vol. XL, p. 463). As a matter of fact; however, a feeble current passes if the water in the voltameter contains dissolved oxygen, or the platinum plates occluded hydrogen.

According to D. Tommasi, "in order that decomposition may take place when a current passes through several electrolytes, it is necessary that the quantity of heat should be equal to the sum of the quantities absorbed by each electrolyte, plus the quantity necessary to overcome the total resistance of the electrolytes. By heat produced by the battery is meant that transmissible to the circuit. In many cases in which there is no decomposition when both electrodes are of platinum, decomposition takes place when the anode consists of some oxidable metal, such as copper or tin." "Of two compounds, that one is decomposed by preference which requires the least thermic energy" (*Journal of the Chemical Society*, Vol. XLII., 1882, pp. 134, 353, 789, 1,019, 1,155, and 1,156; *see* also Favre, "Watt's Dictionary of Chemistry," Vol. VII., p. 458).

On the subject of "The Limits of Electrolysis" consult Berthelot (*Journal of the Chemical Society*, Vol. XLII., 1882, pp. 260 and 353).

According to E. Obach, liquid mixtures of metals do not suffer electrolysis. His experiments were made with alloys of sodium and mercury, of sodium and potassium, and of tin with lead. The portions of alloy around the poles after passage of the current were unaltered in chemical composition (*Journal of the Chemical Society*, 1876, Part II., p. 37 ; *Chemisches Central Blatt*, 1875, p. 497).

Conduction in Electrolytes without Decomposition.—Whether this takes place or not is an important question, and after many researches on the subject electricians are even now not unanimous respecting it. According to Favre's experiments (*Comptes Rendus Académie des Sciences*, Vol. LXXIII., p. 1,463), true conduction without electrolysis does occur when two Smee cells are used to electrolyse dilute sulphuric acid. In the electrolysis of fused argentic fluoride, also with sheet silver electrodes, I observed that the liquid conducted the current with a most extraordinary degree of facility apparently out of all proportion to the weight of metal deposited. Conclusive experimental evidence is, however, still much required to settle the question. With a current of insufficient electromotive force to decompose an electrolyte, either the electric charges must accumulate on the electrodes, and the liquid act as a dielectric, or they must be transmitted by convection or conduction.

Circumstances which Affect the Kind of Deposit.—Both the chemical composition and the physical quality of the substances set free at the electrodes are affected by various circumstances; by the composition of the liquid and its degree of fluidity ; by the strength of the current ; by its density, or strength in relation to amount of surface of the electrode ; by temperature, &c., and by various other circumstances.

The products of electrolysis vary also according to the kind of electrolytic arrangement employed. In the simple immersion one they are mixed with those of voltaic action. In the case of two metals in two liquids separated by a porous division, the products of voltaic action and electrolysis are largely kept separate, and in the case cf electrolysis in an undivided cell by means of a separate current, the anode and cathode products become mixed.

The composition of the liquid is a fundamental condition, and variation of it has usually very powerful effects. The addition of an extra ingredient may cause entirely different substances to appear at each of the electrodes, or alter both the quantity and physical condition of the deposits. An alteration of the degree of fluidity acts similarly, but less powerfully. Whilst also with one strength or degree of density of current a single substance only may appear at each electrode, with a current of greater strength or density additional bodies not unfrequently are liberated. By either decreasing the proportion of water mixed with potassic hydrate, or increasing the strength or degree of density of the current, instead of oxygen and hydrogen alone being evolved, potassium is also set free. A weak current passing through an ordinary silver-plating liquid containing much free potassic cyanide deposits hydrogen only ; but by increasing the density of the current silver is also liberated. It was by obeying these conditions that Davy isolated potassium, and Bunsen deposited chromium. Other investigators also succeeded in obtaining highly oxidable metals in the form of amalgams, without the use of powerful currents, by employing as a cathode mercury, which absorbed the deposits, and thus largely prevented them from redissolving.

The density of the current affects also the physical properties of deposited metals. With a weak current and slow action metals are not unfrequently deposited in a crystalline state, whilst with a strong one they are thrown down as a soft black powder. A nearly saturated solution of cupric sulphate, acidulated with dilute sulphuric acid to a suitable extent, yields ductile metal when the rate of deposition is about half an ounce of metal per square foot of cathode surface per hour. The degree of density of the current not only affects the physical properties of cations, but also those of anions in some cases. Thus a stronger current is usually required to liberate ozone than to set free ordinary oxygen.

Every different metallic solution, and at every different temperature, must be electrolysed at a particular rate in order to obtain from it metal in the state of crystals, reguline metal, or black powder. Some solutions will only yield coherent metal whilst they are hot. If, also, the surface which receives the deposit varies in degree of smoothness, the physical character of the metal is affected. With a viscous solution the quality of the deposit soon changes, because the exhausted layer of liquid next the cathode is only very slowly replaced by solution containing a sufficient proportion of metal.

Different metals whilst depositing exhibit very different physical properties. Copper depositing upon the bulb of a thermometer contracts and compresses the glass bulb, and, causes the mercury to rise. This phenomenon has been termed "electro-striction." Grey metallic antimony depositing very slowly until it has attained 1-10th of an inch in thickness from a solution of tartar-emetic will often crack and curl up in most fantastic shapes. Nickel when deposited to a thickness of half an inch is in the form of smooth, round knobs ; copper has a somewhat similar structure when deposited from certain liquids, the knobs, however, being usually rough.

Circumstances which Affect the Amount of Electro-Chemical Action.—*First,* and essentially, the amount of electro-chemical effect with each substance is strictly proportional to the quantity of current ; double the quantity of current liberates at the cathode double the amount of gas or metal, or causes double the amount of metal at the anode to dissolve or gas be evolved. *Second,* with different substances it varies as their chemical equivalents, or, in other terms, it varies with the atomic (or molecular) weight and degree of valency of the substance. Thus, one atomic weight of any monad element, say silver (the atomic weight of which is $= 108$), requires the same quantity of current as one of any other monad element, say chlorine (the atomic weight of which is $35 \cdot 5$) to liberate it. One atomic weight of any dyad element, say oxygen $(= 16)$, requires twice the quantity of one of any monad element; and one of any triad, say antimony $(= 122)$, or gold $(= 196)$, requires three times the quantity to make it electro-lytically dissolve or deposit, and so on, the proportions in all cases being exactly the same as the ordinary chemical equivalents, and these may be found in any book on general inorganic chemistry. By passing an electric current through two liquids in series, one of which yielded by electrolysis pure copper only, and the other pure antimony only, I found that the weight of copper deposited was $31 \cdot 7$ grains, and of antimony $40 \cdot 6$ grains, and this agreed with one atomic weight of triad antimony or $121 \cdot 98$ parts, being the chemical equivalent of $1\frac{1}{2}$ atomic weights, or $95 \cdot 25$ parts of dyad copper. In a series of electrolysis vessels, therefore, containing different liquids, and electrodes of different metals, the chemical work done by the current at any one anode or cathode is exactly equal in value to that done by it at any other of the same series in the same time. In the decomposition of water, therefore, by an electric current we obtain two parts by weight of hydrogen for each 16 parts by weight of oxygen, and in consequence of the specific gravity of the latter gas being sixteen times greater than that of the former, the relative volumes of them are as two to one. This exact relation of the quantity of the current to the amount of its chemical effect with different substances is known as the law of definite electro-chemical action, and was discovered by Faraday,

With nearly all, if not all electrolytes, Faraday's law of definite electro-chemical action is supposed to be true for even the very smallest currents. The true electro-chemical equivalent of a single substance, however, is only obtained in certain cases. In some instances a portion of the current passes through another ingredient of the liquid, and two substances are deposited simultaneously, and form the equivalent. The alteration of weight of either electrode during electrolysis is often not a true measure of current, because the metal is liable to ordinary chemical action. The true measure is the total amount of substances liberated, taken before they have had time to suffer ordinary chemical change.

Influence of Temperature, &c., on Electrolysis.—Change of temperature has a great effect. Both the electric conductivity and the diffusive power of saline solutions increase by rise of temperature, and each of these circumstances greatly promotes electrolysis. Rise of temperature affects also the relative proportions of current conveyed by the different ingredients of a mixed electrolyte; for instance, I found that in the electrolysis of an acidified solution of cupric sulphate with copper electrodes a considerable deficiency of deposited copper, sometimes amounting to as much as 16 per cent., may result through employing a *hot* solution. ("Electrolysis of Sulphate of Copper," *Proceedings of the Birmingham Philosophical Society*, Vol. III., p. 75; *The Electrician*, Vol. VIII., pp. 271—280). Very few experiments have as yet been made on the influence of great pressure (*see* Section 48) or of magnetism on electrolysis. Remsen, however, found by depositing copper from a solution of cupric sulphate contained in a thin vessel of sheet iron, placed upon the pole of a powerful permanent magnet, that the deposit occurred in a fairly uniform way on the entire surface of the iron except at the parts marking the outlines of the poles. These lines were strongly marked as depressions in the copper. The action was still more striking when an electro-magnet was used instead of the permanent one. In a narrow space marking the outline of the pole there was no deposit. Within this line it was fairly uniform, but outside of it the copper aggregated in irregular ridges, running at right angles to the lines of force, and apparently coinciding with those marking the equipotential surfaces.

Relations of Electro-Chemical to Ordinary Chemical Action.—Electrolytic changes obey the same law of equivalence of action as ordinary chemical ones, and electro-chemical action by a separate current may be viewed as ordinary chemical action taking place in one large and measurable circuit instead of in a multitude of excessively small and non-measurable ones ; and conversely, ordinary chemical corrosion of metals in electrolytes may be viewed as electro-chemical action taking place in an infinite number of such minute circuits.

The electrolytic circuits in which electric currents flow may be of any degree of magnitude, from these small ones upwards, and such currents, of various degrees of magnitude, may circulate simultaneously in the same metals and liquid. Electro-chemical action, therefore, does not necessarily exclude ordinary chemical change. One large current may flow through the electrodes and liquid of an electrolytic cell, whilst "local" action (*i.e.*, ordinary chemical action in patches) in lesser circuits is taking place upon each of the electrodes, and also whilst ordinary chemical action is occurring uniformly upon them.

When an electric current is passed through an electrolyte, whether attended by corrosion of the anode or deposition of metal upon the cathode or not, the layers of liquid in contact with each electrode become changed in chemical composition and density, and are thus indirectly set in motion by the influence of the current, and in consequence of this the ordinary chemical action upon them is altered. In some cases alkali collects around the cathode, and acid around the anode ; in others the liquid around the former becomes more dilute and ascends, whilst that around the latter becomes more saturated with salt and descends. In others, again, insoluble gas is evolved from each electrode, and causes an upward motion of liquid, and in others the gas dissolves in the liquid and alters its degree of corrosive power, as well as of specific gravity ; and in some of these cases, by increasing the density of the current up to a certain point the ordinary chemical action of the liquid upon the cathode is diminished until all such corrosion ceases. This point I have termed "the electrolytic balance of chemical corrosion," and have investigated it in the case of silver in an ordinary cyanide plating solution (see *Proceedings of the Birmingham Philosophical Society*, Vol. III., pp. 268—305, also an abstract in *The Electrician*, Vol. X., p. 381). Many other cases in which ordinary chemical corrosion is balanced and prevented by electro-chemical action remain to be investigated. In some cases the rate of corrosion of a cathode is increased during electrolysis, in consequence of the evolution of hydrogen and consequent motion of the liquid bringing fresh corrosive particles into contact with it (*see* "Corrosion of Cathodes," *Proceedings of the Birmingham Philosophical Society*, Vol. III., p. 305 ; *The Electrician*, Vol. XL, p. 213).

Electrolytic Balance of Chemical Corrosion.— Any mixed electrolyte with a current passing through it and setting free one only of its constituents at a corrodible cathode, and the current then gradually increased until a second cation constituent just begins to be deposited, constitutes an example of "electrolytic balance." Such a case is that of an acidified solution of cupric sulphate, with copper electrodes, with the current increased until metal begins to be deposited ; or that of the ordinary cyanide of silver plating solution containing much free potassic cyanide with electrodes of silver, similarly treated.

Such a silver solution, suffering electrolysis at its "balance point," forms an excellent illustration of the compensation and balance of a number of molecular forces, the alteration of any one of which disturbs the remaining ones. Even a change of temperature may be included in this statement. It is a case of balance of powers in which the state of equipoise depends upon the united and simultaneous action of at least seven or eight different influences, viz., ordinary chemical corrosion, strength of current, nature of cathode, size of cathode, temperature, proportions of water, of argento-potassic cyanide, free potassic cyanide, and of the soluble salts, &c., present in the form of impurities. Additional causes, or conditions, might also be introduced, which by their presence would probably affect the state of balance, such, for instance, as by dissolving in the liquid various salts or other substances.

Several of these causes or conditions may be modified so as to alter the balance either in one or the opposite manner. Thus, increased chemical corrosive power, a larger cathode, more free potassic cyanide, less argentic cyanide, or less strength of current alter the balance in one direction, whilst their opposites alter it in the contrary one.

Such experiments also show that the various conditions of the state of balance may either assist or counteract each other ; that an increase of current is equivalent to a decrease of argentic cyanide, if the one is increased the other must be decreased in order to maintain the state of the balance ; that an addition to the amount of free potassic cyanide, by diminishing resistance, is equivalent to an increase of current ; that a decrease of cathode surface necessitates either a decrease of argentic cyanide or of current; that a rise of temperature is balanced by an increase of current, and so on.

All these influences have numerical values. In the experiments referred to it is shown that a rise of temperature of the liquid of 60 Fahrenheit degrees, *i.e.*, from 60° to 120° F., is balanced by an increased strength of current from ·002306 to ·003282, or ·000976 ampere.

The arrangement and use of a depositing solution in such a manner constitutes a method of detecting the molecular influences of substances dissolved in electrolytes, and of determining to a certain extent their kind and amount of influences by their effect and degree of power in altering the "balance point" either in one direction or the opposite. It was found that the mere presence and admixture in solution of even a small quantity of argento-potassic cyanide in the above liquid altered the molecular arrangement of the free potassic cyanide in such a way as to diminish its power of alone transmitting the current into a silver cathode, and increased the tendency of the current to pass into the cathode partly by means of the double salt.

The phenomena of the "balance point" constitute also an interesting example of molecular equilibrium, in which the "balance point" may be compared to a ball suspended by an elastic cord and having attached to it a number of other similar cords in a state of tension, each drawing it in a different direction. In such a case an alteration of the degree of strain of any one of the cords changes that of all the others, and alters the position of the ball.

Secondary Effects of Electrolysis.—In very many cases the new substances actually observed at the electrodes are not those set free by the current, but are products or results of the action of those substances upon the liquid or upon the electrodes. Thus, when potassium is deposited from a solution of one of its salts into a cathode of mercury, the liquid in contact with the mercury becomes alkaline; when iodine is set free from the cathode in a solution of iodic acid, it is due to the deoxidising action upon the iodic acid of the hydrogen liberated there by the electrolysis of the water or of the iodic acid, and when it is set free at the anode during electrolysis of a solution of hydriodic acid it may be viewed as a direct result of the current or as a secondary result of liberated oxygen. The peroxide of silver formed at an anode of platinum in a solution of argentic nitrate may be viewed as a secondary product due to the action of the liberated oxygen or ozone upon the silver of the liquid. In many cases it is difficult to determine whether a liberated substance is due to primary or to secondary action.

Faraday advanced the view that "only those compounds of the first order are *directly* decomposable by the electric current which contain one atom of one of their elements for each atom of the other ; for instance, compounds containing one atom of hydrogen or metal with one atom of oxygen, iodine, bromine, chlorine, fluorine, or cyanogen, whilst boracic anhydride (BO_3), sulphurous anhydride (SO_2), sulphuric anhydride (SO_3), iodide of sulphur, the chlorides of phosphorus (PCl_3 and PCl_5), chloride of sulphur (S_2Cl), chloride of carbon (C_4Cl_6), tetrachloride of tin ($SnCl_4$), terchloride of arsenic ($AsCl_3$), pentachloride of antimony ($SbCl_5$), are non-conductors of electricity, and incapable of electrolysis." I have observed that the decomposability of a salt depends upon the kind of liquid in which it is dissolved ; *e.g.*, the iodide and bromide of antimony, both of which conduct and are decomposed when dissolved in acidulated water, do not conduct and are not decomposed when dissolved in carbonic bisulphide.

Insoluble Coatings on Anodes.—In many cases of electrolysis of aqueous solutions, the anode does not dissolve, but becomes coated with an oxide, chloride, fluoride, cyanide, sulphate, or other insoluble salt, usually by chemical union of the metal with an ingredient of the liquid. In this way silver in dilute hydrochloric acid becomes coated with argentic chloride ; in a solution of argento-cyanide of potassium it becomes covered with argentic cyanide, lead in dilute hydrofluoric acid becomes coated with fluoride, and so on. In some cases the insoluble coating occurs not by corrosion of the anode, but by the oxygen evolved by electrolysis of the water acting upon the ingredients of the solution ; in this way various peroxides are formed. The formation of peroxides occurs upon platinum anodes in solutions of the nitrates of bismuth, silver, and lead ; in certain alkaline solutions of lead, nickel, and cobalt, and in those of nitrate and acetate of manganese, and when the films which are thus formed are exceedingly thin their colours are in some cases very magnificent.

Electrolytic Alloys.—In many cases when metals are deposited upon metals, the two substances form alloys ; grey antimony deposited upon mercury from a solution of tartar emetic alloys readily, but the black explosive variety does not. Tellurium deposited from a solution of its chloride upon platinum also forms an alloy. Boron, silicon, and lithium, when deposited from certain fused compounds upon a surface of platinum, also alloy with it. Hydrogen deposited upon palladium, or upon certain other metals, iron in particular, is absorbed, and imparts to the metal peculiar properties. In some cases the absorption of the deposited substance continues after deposition has ceased ; this is only visible in cases where the deposited coating is extremely thin. A thin film of deposited copper is absorbed by zinc.

Purity of Electrolytic Deposits.—Not only do the deposited substances sometimes alloy with or penetrate into the mass of the cathode, but in some cases during the act of deposition they combine with some of the elements of the electrolyte, and are thereby altered in property. In this way antimony which has been rapidly deposited from a strongly acidified solution of its oxide in hydrochloric acid contains several per cent. of the salt derived from the liquid, and possesses the very remarkable property that, if broken, or even scratched, it suddenly rises in temperature about six hundred Fahrenheit degrees ; it also has the appearance of highly burnished steel, very widely different from the colour and appearance of the pure grey metal very slowly deposited from a feebly acidified solution of tartar emetic in dilute hydrochloric acid. This black antimony gradually loses its latent heat, explosive power, and brilliant appearance, in the course of one or two years, the period varying according to the thickness of the deposit.

It is only in certain cases and in the presence of a collection of suitable fortuitous conditions that a deposited substance is extremely pure ; substances very easily deposited, such as hydrogen and copper, are usually so chiefly because it requires a stronger power to deposit most other bodies, also because in some cases the other easily deposited metals are precipitated as insoluble salts by ordinary chemical action. Thus lead in an acidified solution of sulphate of copper is precipitated as sulphate ; similarly silver is precipitated in a solution containing a dissolved chloride. All deposited substances are, of course, more likely to be pure the greater the degree of purity of the liquid.

Divided Electrolysis.—When an impure liquid or a mixture of solutions is electrolysed, either a single substance alone may appear at the anode or cathode, or several may be simultaneously liberated. With a feeble current and large electrodes one substance alone may appear at either electrode, but by either increasing the strength of the current or diminishing the size of the electrodes, a second, or even a third substance may be liberated, the current appearing to divide its action amongst the various compounds present. The least electro-positive cation is usually liberated first, and the more positive ones subsequently as the current strength is increased. In all cases weaker affinities appear to be overcome first ; but this is only a superficial explanation, the true one being much less simple. By employing proportions of the substances, larger as their electro-positive property in the particular liquid is greater, several may be simultaneously deposited or the more positive ones may be deposited even in larger amount than the less positive ones, as for instance potassium from moist potassic hydrate. It is by obeying these and other conditions that alloys and mixtures of substances are usually set free at the electrodes. The order of degree of electro-positive state of the metals desired to be deposited may in most cases be ascertained by connecting the metals in pairs with a galvanometer, immersing their free ends in the liquid, and observing the direction of deflection of the needles.

In electrolysing a mixture of the sulphates of zinc, cadmium, and copper, Favre succeeded, by altering the conditions of the experiments, in obtaining at will either one, two, or all the three metals simultaneously ; and states that the results of the operation vary, 1st, with the voltaic energy of the battery ; 2nd, with the electrolytic resistance of the salts ; 3rd, with the relative quantity of each salt ; and 4th, with the greater or less rapidity of the electrolysis, which can be regulated. He concludes that by varying these conditions we are enabled to withdraw from a mixture of salts the different metals in succession, and thus proposes an electro-chemical analysis (*Comptes Rendus*, Vol. LXXIII. ; *Journal Chemical Society*, 2nd series, Vol. X., p. 113). This proposal, however, is not a new one.

Polarisation of Electrodes.—In consequence of the alteration both of the chemical composition of the surface of the anode and of that of the cathode, and also of that of the layer of liquid in contact with each of the electrodes by electrolytic action, the electric state of each of these surfaces is continually liable to change ; or, in other words, the surfaces become polarised. And as the substances set free at the anode are usually electro-negative, and those at the cathode are usually electro-positive, the electric states produced by polarisation are opposite in kind to the original ones, and tend to produce an electric current in an opposite direction to the previous one, and therefore weaken that current. According to M'Gregor (*Nature*, July 19, 1883, p. 283), the degree of polarisation of electrodes is independent of their degree of difference of potential. By passing an undivided current by means of four similar platinum sheet electrodes through two cells containing equal sections but unequal lengths of dilute sulphuric acid (the current being therefore of equal density in each and the electrodes of the two vessels of unequal potential), he found that the variation by lapse of time of the electromotive forces of the two cells after cessation of the polarising current was similar.

Unequal Electric Action at Electrodes.—By the electrolysis of a metallic electrolyte by means of vertical corrodible electrodes, the liquid around the anode usually becomes more saturated with metallic salt, and being heavier descends, whilst that around the cathode becomes deprived of metal, acquires less specific gravity, ascends, and spreads itself over the surface. In consequence of these variations in specific gravity of the upper and lower parts of the electrolyte, the direction of the current in it is gradually affected. At first, whilst the liquid is uniform in density and composition throughout, the whole of the current is perfectly horizontal in direction, and equal amounts of it pass through equal sections of the liquid ; but, if the current is sufficiently strong, after a while it passes unequally, and the bulk of it travels in an oblique direction from the upper part of the anode to the lower part of the cathode. In consequence of this, the greatest amount of electrolytic effect is at those parts of the electrodes, and thus in some cases the upper part of the anode is rapidly corroded, whilst the lower part of the cathode receives a rapid deposit. If, however, the current is very feeble, the liquid is kept uniform in composition by means of diffusion as fast as it is rendered non-homogeneous by electrolysis ; if also the liquid is very viscous and diffusion difficult, these phenomena are more slowly produced. These changes in composition of the upper and lower parts of the liquid also give rise to local currents, which leave the upper part of each electrode and re-enter at its lower portion, and produce the usual electrolytic effects. With the electrodes horizontal, and the cathode below the anode, the above inequalities of electrolytic action do not. occur.

In some cases, apparently in consequence of a very thin layer of corrosive liquid collecting upon the surface of the electrolyte by long-continued rapid electrolysis, the anode is gradually cut off at that level and falls to the bottom; in other cases, partly in consequence of evaporation, of oxidation of moist metallic surfaces by the air, and of capillary action, the cathode is corroded in numerous short vertical grooves at the surface line of the liquid, and a narrow line of metallic deposit gradually forms above the surface of the liquid, and follows the outline of the narrow film of liquid which has risen by capillary action. A probably correct explanation of the formation of this deposit is that the capillary film of liquid becomes much less corrosive and more saturated with the metal by chemical action than the bulk of the liquid beneath. The piece of metal is therefore in contact with two liquids of different chemical composition, and a voltaic element is thereby formed and generates a current, the positive electricity of which proceeds from the portion of surface of metal which is in the upper and nearest part of the bulk of the solution into that solution, thence to the capillary film, and into the narrow surface of metal in immediate contact with it, and thus corrodes the metal just below the surface of the liquid, and deposits the metal in the capillary film. This phenomenon is seen in alkaline liquids as well as in acid ones; for instance, with silver in a solution of potassic cyanide.

The corrosive effect attending this capillary action differs somewhat with different metals and liquids. With metallic tin, in particular, in dilute hydrochloric acid, in some experiments of mine, grooves about ·5mm. deep were corroded in its surface and extended in a vertical direction to a distance of nearly 7mm. above the level of the liquid. The grooves were crooked, and had branches like those of a tree, and those upon the cathode were longer and deeper than those on a similar sheet of metal in a separate portion of the same liquid not under electrolysis.

Dependence of Electrolysis upon Liquid Diffusion.—This is a branch of the subject which has hitherto been but little examined, and much remains to be discovered in it. Many of the phenomena of electrolysis are, no doubt, essentially related to the power of liquid diffusion. An extremely viscous liquid admits of but slow electrolysis. Long has discovered (*Phil. Mag.*, 1880, Vol. IX., p. 425) that in almost every case the best conducting saline electrolytes are solutions of those salts which have the fastest rate of diffusion, and those are usually the salts which have the largest molecular volume, and which also absorb most heat in dissolving. He also arrives at the conclusion that "the rate of diffusion of a salt is proportional to the sum of the velocities with which its component atoms move during electrolysis."

Electrolytic Diffusion of Liquids.—I have experimentally investigated this converse part of the subject (see *Proceedings of Royal Society*, No. 203, 1880, p. 322, and No. 212, 1881, pp. 56—84), and have shown that an electric current will cause a liquid to diffuse, and I discovered that when such a current was passed up or down through the surfaces of mutual contact of certain aqueous solutions of different specific gravities lying upon each other in well-defined layers, the bounding surfaces of contact of the two liquids became indefinite where the current passed downwards from the lighter to the heavier solution, and became more sharply defined where it passed upwards from the heavier into the lighter one; and that, on reversing the current several times in succession, after suitable intervals of time, these effects were reversed with each such change of direction; also, in various cases in which the contiguous boundary films of the two liquids had become mixed, and the line of separation indefinite, the liquids separated by the influence of the upward electric current, and the line of separation became as perfect as that between strata of oil and water lying upon each other. I also observed, 1st, the production of definite lines, not only where the current passed from the heavier into the lighter solution, but also (in certain cases) at the surface where it passed from the lighter to the heavier one. 2nd. The production in some cases of two or three separate lines at the former situation, and less frequently also at the latter one. And, 3rd, an apparent movement of the mass of the heavier solution, usually in the direction of the electric current, but in certain exceptional cases in the reverse direction. By further experiment (see *Proceedings of the Royal Society*, No. 217, 1881, p. 141) I ascertained that, 1st, in certain cases the upper and lighter liquid diffused downwards continuously through the meniscus, or surface of separation of the two liquids, during the passage of an upward electric current; and, 2nd, that during the continuance of the current either no manifest expansion of the upper liquid occurred, and that equal volumes of liquid diffused in two opposite directions through the meniscus, or that any expansion of the upper liquid was compensated by downward diffusion of an equal bulk of that liquid; or that the united volumes of metal deposited from the upper liquid, and of the acid element from which it had been separated by electrolysis, were greater than before such separation, and that this was compensated by the volume of liquid diffused downwards through the meniscus. In these latter experiments the meniscus retained its position during the passage of the current, thereby proving that the actual bulk of the upper liquid remained the same whilst diffusion of a portion of that liquid took place downwards through the meniscus.

Transport of Ions.—Hittorf and G. Wiedemann found that usually the velocity of transport in electrolysis of anion and cation are different, and F. Kohlrausch discovered that in dilute solutions of salts, acids, and alkalies every ion under the influence of currents of equal density moves with its own particular velocity, independently of others moving at the same time in the same or opposite direction. The order of velocity of cations, the first named being the fastest, was hydrogen, potassium, ammonium, silver, sodium, barium, copper, strontium, calcium, magnesium, zinc, lithium; and of anions was hydroxyl, iodine, bromine, cyanogen, chlorine, NO_3, ClO_3, and the halogen of acetic acid.

Relations of Electrolysis to Heat.—In consequence of chemical action and of the passage of an electric current from one substance to another, changes of temperature occur at each electrode, and at each junction of two different liquids. These changes are different in every different case, and have been but little investigated. With an anode of copper in an acidulated solution of its sulphate, heat is evolved by the oxidation of the metal; but with one of platinum in dilute sulphuric acid, heat is absorbed, and oxygen is reduced to the elementary state. At the cathode, in the former liquid, copper is liberated and heat absorbed; but with a platinum cathode in nitric acid, heat is set free by oxidation of the deposited hydrogen.

According to Favre (*Comptes Rendus*, Vol. LXXIII., pp. 1,036—1,085, 1,186—1,262), although in certain cases the metal dissolved at the anode is all reproduced at the cathode, heat is liberated which is not transmissible to the circuit. The oxides and salts of the alkali metals, when subjected to electrolysis, are decomposed, and give up their metal, which metal being *directly oxidised* at the expense of the water, sets free a quantity of heat which reinforces the voltaic energy of the battery. The secondary reactions which accompany electrolysis and produce heat not transmissible to the circuit always tend to strengthen the energy of the battery whenever the current is weak and when the electrolysis offers great resistance. Such secondary reactions are, for example, produced by the hydrogen and the oxygen set free during electrolysis, the first being burned, the second oxidising any oxidable substance present (*Journal of the Chemical Society*, 2nd series, Vol. X., pp. 110—113).

In addition to changes of temperature produced by electro-chemical and chemical actions in the electrolyte, heat is evolved by conduction-resistance in the mass of the liquid; and I have noticed that if two large masses of the same, or of two different electrolytes, are united by an open short glass tube of the shape of an hour-glass, and of small diameter, by employing a sufficiently strong current the liquid in the narrow part of the connecting tube may be caused to boil ("Influence of Voltaic Currents on Diffusion of Liquids," *Proceedings of the Royal Society*, 1881, No. 213, pp. 76—82).

Gladstone and Tribe have shown by experiment that if a strip of metal is immersed at its two ends in a salt of the same metal in a state of fusion, but of unequal temperature at the two parts where the metal dips into it, the hotter end of the metal dissolves, and the less heated part receives a metallic deposit. Copper in fused cupric chloride is an example (*Journal Chemical Society*, Vol. XL., 1881, p. 868).

Theories of Electro-Chemical Action.—Various theories have from time to time been proposed to account for the leading phenomena of electrolysis, but none of them have as yet been very clear or satisfactory. One of the best is that propounded by Faraday. He considers that electrolysis resulted from a peculiar corpuscu'ar action developed in the direction of the current; and that it proceeded from a force which was either added to the affinity of the bodies present, or determined the direction of that force. That the electrolyte was a mass of acting particles, of which all that were in the course of the current contributed to the terminal action, and in consequence of the affinity between the elements being weakened, or partially neutralised by the current parallel to its own course in one direction, and strengthened and assisted in the other, the combined particles acquired a tendency to move in different directions. The particles of one element, a, cannot travel from one pole to the other, unless they meet with particles of an opposed substance, b, ready to move in the opposite direction. For in consequence of their increased affinity for these particles, and the diminution of their affinity for those which they have left behind, they are continually driven forward.

Any tolerably complete theory of electrolysis of a fundamental character must, however, be a mechanical one, based upon the assumption of molecular motion, and expressible in mathematical and geometrical terms. Whilst, also, the theory must represent the kind of molecular motion which constitutes an electric current, it must also be consistent with the numerous and varied phenomena attending electrochemical action. And as the essential kinds of molecular change which occur at the electrodes are probably more or less modified in every different case, a complete theory must admit of varied application. Clausius considers that the atoms or groups of atoms forming a molecule of an electrolyte revolve around one another, similarly to planets, and are sometimes nearer to and sometimes farther from each other ("Poggendorff's Annalen," CLVI., pp. 618 to 626). Favre states (*Comptes Rendus*, Vol. LXXIII., p. 971) that in each voltaic couple the molecules are electrolysed *successively*, and that when the *absolute number* of vibrations which correspond to a given intensity of the current have been determined the *absolute weight* of the chemical molecules will be known (*Journal Chemical Society*, 2nd series, Vol. X., p. 25).

The immediate or primary electrolytic changes are evidently a result of molecular energy transmitted along the wires from the source of the current; and the energy so transmitted is substantially the same in its chief properties and electrolytic effects, whether it proceeds from a voltaic battery, a thermopile, or a dynamo electric machine. Any theory, therefore, which explains electrolysis must also be consistent with the fact that in the act of electrolysis the homogeneous electric energy is converted into potential molecular energy as varied in kind as the properties of the liberated elements. It must also explain why the same element may in certain cases be an anion in one combination and a cation in another.

Distinction between Voltaic and Electrolytic Action.— These two actions are almost entirely the converse of each

other ; the former is a consumer, and the latter a producer of
potential molecular energy. In voltaic action substances are
burned, in electrolytic they are unburned. In a voltaic cell
potential or stored-up energy of elementary substances is converted into electric current ; in an electrolysis vessel current
is converted into stored-up potential energy in the elementary
substances liberated at the poles.

Intimate Connection of Voltaic and Electrolytic Action.—
As in nearly every voltaic circuit the current produced at the.
positive surface decomposes the liquid at the negative one, and
in nearly every electrolytic circuit voltaic currents are produced by difference of chemical composition of the liquids in
contact with the two electrodes, nearly every voltaic circuit is
partly electrolytic, and nearly every electrolytic circuit is
partly voltaic.

According to these views, voltaic action is chemico-electric,
and a case of chemical union in all cases ; and true electrolytic
action is always a case of electro-chemical separation, sometimes accompanied by chemical union at the electrodes.

The various phenomena of electrolysis are produced not
only by electric currents proceeding from an external source,
but also by those produced in the electrolyte itself ; and also
not only by currents generated and flowing in circuits of
measurable magnitude in that liquid, but also by others in
circuits so small that they cannot be measured.

In the case of an ordinary voltaic cell or electrolytic
vessel, the positive and negative surfaces are sufficiently far
asunder to enable us to perceive the action at each ; but in
those of "local action" and minute circuits, such as those
in cases of deposition by "simple immersion," or the chemical
substitution of one metal for another, as when iron becomes
coated with copper by simply immersing it in a solution
of cupric sulphate, the positive and negative surfaces of
each circuit are so excessively small, so exceedingly near
together, and the circuits are so numerous that they cannot be
separately observed, and the entire immersed surface of the
metal is covered with inseparable voltaic and electrolytic actions.

The substances set free by electrolysis do not always
appear ; the instant they are liberated they are subject to
ordinary chemical action by contact with the liquid, the
electrodes, and the atmosphere. Thus, when potassium is
set free at the cathode from a solution of any of its salts, it
is instantly oxidised into potash ; or oxygen set free at a
copper anode instantly oxidises the copper. Other relations
of electrolytic to ordinary chemical action have already been
described.

These facts show the intimate connection of chemical,
electro-chemical, and voltaic phenomena ; that the study of

electro-chemistry requires considerable knowledge of voltaic electricity; and that the modes of electrolysis require to be classified according to the magnitude of the electric circuits and the degree of complexity of the voltaic and electrolytic combinations employed. Neither voltaic action nor electrolysis can be successfully studied without also a previous knowledge of general chemistry. As the subject of these articles is electrolysis and not voltaic action, the latter will only be explained so far as is necessary to elucidate the former.

Modes of Generation of Voltaic Currents.—A voltaic current may arise—First, from the contact of two metals with one liquid, e.g., zinc and copper in dilute sulphuric acid; second, from the contact of one metal with two liquids, e.g., two pieces of silver, one in a solution of potassic cyanide, and the other in argento-cyanide of potassium, the two liquids touching each other through an intervening porous partition, or by lying upon each other; or third, from the contact of two metals with two liquids so arranged, e.g., zinc in dilute sulphuric acid, and copper in a solution of cupric sulphate.

The strength of current thus obtained is usually the greater the more wide the difference in the chemical properties of the metals and liquids employed, and is commonly the greatest with the combinations of two metals with two liquids.

Source of the Current.—Theory of Voltaic Action.—Two rival theories of the source of the current have long been entertained—First, that of Volta, that the current is due to contact of dissimilar conductors of electricity; and second, that of Faraday and other English investigators, that it is due to chemical action. Neither of these views, however, is completely satisfactory, or has been universally accepted.

If, however, we adopt a theory that *the molecules of substances* (those of chemically active bodies in particular) *are in a state of ceaseless motion* (that of frictonless bodies in a frictionless medium, the universal ether) *until they chemically unite*, an efficient cause of the current (and of chemical action) becomes at once exceedingly clear.

. According to this view, which I may term the Ceaseless Molecular Motion Theory of voltaic and chemical action, neither contact nor chemical action is the real dynamic cause of the current, but the true cause is the potential molecular energy of the corroded metal, and of the corroding element of the liquid with which it subsequently unites, and chemical corrosion is only the *process* or *mode* by which the molecular motions of those substances are transformed into heat and current.

Both the heat and electric current produced during the chemical corrosion of metals by electrolytes are recognised modes of motion, or forms of active molecular energy, and as motion or energy cannot be created, but can only result from the expenditure of some other form of motion, these movements are derived from the original metal and liquid, and the corroded metal and liquid employed have, after the action, lost to a greater or less extent their power of further producing heat or current.

According to this view, also, contact is only *a static condition* which enables the molecular motions of the one substance to modify those of the other, and thus produce static electric polarity; and this, if sufficiently strong, produces corrosion and the new modes of motion, namely, heat and current.

Electrical Theory of Chemistry.—This theory (attributed to Berzelius) assumes that the chemical union of any two substances is an electrical act; *i.e.*, that during contact, previous to union, the one substance is relatively positive, and the other relatively negative, and that the act of union is a consequence of these states; also that during the act of union the two electric states neutralise each other and produce heat and current.

In accordance with this theory, and with the voltaic series of metals, the various elementary substances have been arranged in the following order, the most strongly electropositive substance being placed first, and the most negative one last:—Caesium, rubidium, potassium, sodium, lithium, barium, strontium, calcium, magnesium, aluminium, zinc, cadmium, iron, cobalt, nickel, lead, tin, copper, mercury, silver, palladium, gold, iridium, rhodium, platinum, hydrogen, osmium, antimony, tellurium, arsenic, silicon, carbon, phosphorus, selenium, iodine, bromine, chlorine, nitrogen, sulphur, fluorine, oxygen.

The electrical theory of chemical action may be reasonably extended from that of metals and electrolytes to that of all non-conducting elements in non-conducting liquids, because resistance to conduction is only of degree, and not infinite. If, therefore, the electric polarity produced by the molecular motions of bodies is sufficiently strong, and the electrical circuits sufficiently small, chemical union and electrolysis in non-conductors must occur.

The deposition of copper and silver from aqueous solutions of their salts by immersing in them a piece of ordinary phosphorus, are good examples of electrolysis produced by a non-conducting element in conducting solutions, and the separation of hydrogen from pure water by contact of a zinc-platinum couple is an instance of electrolysis by conducting bodies in a non-conducting liquid. And the chemical decomposition of non-conducting liquids by non-conducting elements may be regarded as only an extension of the same kind of action.

Voltaic Series.—The degree of power of generating a voltaic current differs with every different metal and liquid.

The relative power of two metals is usually ascertained by connecting them with the ends of a galvanometer coil, then immersing the free ends of the metals simultaneously in the particular liquid, and observing the direction of deflection of the galvanometer needles. The strongest acting metal is electro-positive.

In this way numerous tables of what are termed voltaic series of metals in various electrolytes have been obtained, which differ somewhat with every different liquid, and also with the same liquid of different temperatures or strengths, but are usually approximately in the above order. For the order in any particular liquid the reader is referred to text books on electro-metallurgy, and to special researches on the subject. Extensive series may be found in " Gmelin's Handbook of Chemistry," Vol. I., p. 397 ; also *Proc. Roy Soc.*, No. 200, 1879, pp. 38—49 ; and in 'Electro-Metallurgy," Longman's Text Books of Science.

The above series are only those obtained by immersing two different metals in one liquid; others are obtained by immersing two pieces of the same metal in two different liquids separated by a porous partition ; and more complex ones might also be formed by immersing two metals in two liquids thus separated ; and series may also be obtained by the employment of fused electrolytes in place of the usual aqueous solutions.

Voltaic Batteries.—Voltaic elements are simply combinations selected from series arrived at in the above manner, those being selected which include the best combination of desired qualities, such as strength of current, cheapness of metal and liquid, manageability, freedom from offensive fumes, &c. A strong element can only be obtained by selecting metals which are far asunder in the " series." The strongest are those formed of two metals and two liquids. The varieties of batteries are very numerous, and a complete description of them would fill a volume.

Voltaic Currents.—The continuous union of the two electricities, or electric states of positively and negatively charged substances, through a conductor, constitutes an electric current ; and the chief circumstances to be considered in connection with such currents are polarity, potential, electromotive force, quantity, strength, and density of current, conduction, resistance, polarisation, &c.

Polarity, Potential, Electromotive Force, &c.—Assuming the Ceaseless Molecular Motion Theory of voltaic (and chemical) action to be the correct one, we may consider the pre-existing molecular vibrations of the metals to be the cause of volta-static polarity and all its consequences ; that when two different metals are brought into mutual contact, the molecular motions of the two metals act upon each other ; and the composition of forces causes the one metal to become positive and the other negative ; also that, when a metal is brought into contact with an electrolyte, similar effects of polarity occur.

Previously, therefore, to the completion of the circuit and formation of a current, the two metals, by contact with an electrolyte, become charged with the two kinds of electricity; in a statical condition, and are in a state of electric potential or pressure, capable of doing electric work by their subsequent discharge. This difference of electric potential produces electric flow, like a difference of pressure of water produces a flow of that liquid. The electric charges of the metals are in a state of tension tending to escape, and may be detected by means of an electroscope or measured by an electrometer ; the degree of tension is, however, exceedingly minute. The charged state also produces induction, which acts from molecule to molecule during discharge, and precedes current.

Electromotive force, or the power which moves, or tends to move, electricity from one place to another, varies with every different voltaic couple, and with the same couple at every different temperature ; and these differences may be detected, by opposing the two couples to be compared, in single series in circuit, with their terminals connected to those of a galvanometer ; the current from the strongest then produces a deflection of the needles. In a voltaic series, the metals are arranged, in the order of their relative degrees of electromotive force.

The degree of electromotive force of a couple depends considerably upon the degree of difference of strength of chemical affinity of the two metals for the electro-negative elements of the liquid ; and the farther asunder the metals are in the chemico-electric or volta tension series, the greater usually is the electromotive force of the current they produce. All other circumstances being alike, the most rapidly corroded metal, used with the least corroded one, usually gives the current of greatest electromotive force.

The measurement of the degree of electromotive force of a voltaic cell is usually made by comparing it with that of some convenient and steady source of current, such as that of a Daniell or a Clark cell. The unit of electromotive force (E) is termed a volt, that of a Daniell cell is $= 1\cdot078$ volt, and that of a Clark $= 1\cdot457$ volt. For measuring feeble electromotive forces I have devised a convenient form of thermopile, consisting of about 300 pairs of iron and German silver wires, and have employed it in making a great number of measurements, not much exceeding that of one Daniell. It is capable of measuring differences of $\frac{1}{38000}$th of a volt. (See *Proceedings of the Birm. Phil. Soc.*, Vol. IV., Part 1.)

Resistance.—Every conductor of electricity, no matter how good it may be, is an obstacle to the passage of a current. Electrolytes offer great resistance, especially with anodes composed of a metal which does not readily dissolve in them. Perfectly pure water with platinum electrodes hardly transmits any current from a single voltaic cell. The degree of resistance of a saturated solution of sulphate of copper at 48° F.—and this is a comparatively good conducting electrolyte—is nearly 17 million times that of a copper wire of equal length and section at 32° F. Tables of the conduction resistance of various liquids are contained in most works on voltaic electricity.

According to Quincke ("Pogg. Annalen," Vol. CXLIV., pp. 1–33, 161–190), as long as the density of the current in the liquid is too small to overcome the chemical affinity the liquid will behave as an insulator, but it may become conducting by an increase of that density. Liquids conduct, according to Ohm's law, the same as solids (*Journal Chemical Society*, 2nd series, Vol. X., p. 208).

The total resistance in an electrolytic circuit is usually divided into internal, or that in the battery, and external, or that in the remainder of the circuit; there is resistance in the battery itself, in the liquid, and especially at the surface of the negative plate, if hydrogen is evolved there.

The ordinary unit of resistance (R) is termed an ohm, and is that offered at 0° C. by 1·0486 metre length of mercury of 1 square millimetre section. The amount of resistance in a wire, A, is conveniently measured by dividing the current from a very small Daniell cell, so that one portion shall pass through A and one wire, B, of a differential galvanometer, and the other portion through another wire of known resistance, C, and the other wire, D, of the galvanometer in the opposite direction to that through B, and altering the length of A until the needles of the instrument stay at zero. The resistance in A and C is then equal. The measurement of resistance of an electrolyte is much more difficult on account of the varying polarisation of the plates, but may be effected in a somewhat similar manner by making two measurements by means of a very feeble current after the polarisation has become steady— one when the electrodes are near together, and the other when they are far asunder, using in each case electrodes as large as the transverse section of the liquid, and in certain cases of the same metal as that of the salt of the electrolyte in order to diminish polarisation. The difference of resistance of the two measurements is the amount of resistance of the difference of length of liquid in the two cases.

Strength of Current.—The strength is the amount which flows through any transverse section of the circuit in a given period of time, and the amount flowing at any given instant is the same in every such section of the circuit, whether that section be large or small; the unit of time employed is one second. It varies directly as the electromotive force, and inversely as the total resistance in the circuit (Ohm's law).

A given voltaic cell can only yield a certain maximum strength of current, and any conductor introduced into the circuit diminishes that amount. The greater the electromotive force of a current, the less is it diminished by increase of external resistance; such a current is said to possess "great intensity." If the external resistance is very small, an increase of electromotive force of the battery adds very little to the strength of the current; but if it is large, the opposite effect takes place. The difference of effect produced by means of a current from a single cell, and one from many, does not arise from any difference in the nature of the current in the two cases, but from the difference of proportion of internal to external resistance. No difference has hitherto been proved to exist in any two currents of equal strength.

The unit of strength of current (I) is termed an ampere, and is the strength produced by an electromotive force of 1 volt in a circuit having a resistance of 1 ohm. The strength (or quantity per second) of a current may be measured by passing the current during a known period of time, either by means of platinum electrodes through dilute sulphuric acid in a voltameter, and measuring the evolved hydrogen, or by means of silver electrodes through a solution of argento-cyanide of potassium, containing the minimum practicable amount of free potassic cyanide, and weighing the deposited silver. The latter method gives a little deficiency, owing to a small amount of the current passing through the free cyanide. Each ·000162 grain of hydrogen or ·017343 grain of silver deposited per second equals 1 ampere. Additional methods of measurement are usually described in text books on voltaic electricity.

Unit of Quantity of Current.—Whilst the degree of intensity of chemical action between two substances determines the electromotive force of the current, it is the quantity of substances uniting which determines its amount. The unit of quantity of current (Q) is termed a coulomb; it is one very little used, and is the amount which a strength of one ampere gives in one second. Measured by the method of electrolysis, it is that which deposits ·000162 grain of hydrogen, ·0051035 grain of copper, or ·017343 grain of silver.

Density of Current.—This means merely the strength of current passing through a given section of a conductor, or into or out of a given sized surface of electrode. No unit of it has hitherto been commonly recognised, but I have proposed

(*Proc. of Birm. Phil. Soc.*, Vol. III., p. 277) the unit strength of current entering a surface of one square centimetre of cathode as a convenient one.

Density of current at the surface of the electrodes is one of the most important circumstances in electrolysis. Variation of it has often great effect both upon the physical structure and chemical composition of deposits upon cathodes; the former has already been described. It also appears to affect the properties of oxygen and chlorine when they are separated at the anode. Metals which are easily oxidised, such as cobalt, are deposited upon cathodes in a state of oxide or basic salt if the density of the current at that surface is small, but in the state of metal if it is great. It was largely by increasing the density of the current that Davy succeeded in isolating potassium. Any circumstance, such as polarisation, which diminishes the density of the current, is liable to affect the properties and composition of the deposit.

Quincke has shown that the force tending to separate the elements of an electrolyte is proportional to the strength of the current per unit of sectional area of the liquid; that it increases with the electromotive force of the current, and is inversely proportional to the length, but independent of the cross section and conductivity of the liquid, if the resistance of the remainder of the circuit is small in comparison with that of the electrolyte (*Journal of the Chemical Society*, Vol. X., p. 208).

Distribution of Current in Electrolytes.—With a perfectly homogeneous electrolyte of much larger section than the opposed surfaces of the electrodes, and the latter placed centrally and symmetrically in it, when the current leaves the anode it spreads out in the liquid in curves not unlike those of magnetism diverging from the poles of a magnet, and the densest portion of the current is in the central axis joining the electrodes. Its distribution in the liquid has been investigated by Tribe, who suspended little bits of metal in different parts of a cross section of the solution, and ascertained the amount of electrolytic action produced upon them by the same current, during the same period (*Proceedings Royal Soc.*, Vol. XXXI., p. 320; Vol. XXXII., p. 435).

Relative Amounts of Currents produced by Different Metals.—Equal weights of different metals yield by voltaic action different amounts of current. Whatever amount of current a particular weight of any given metal requires in order to deposit it, that same amount will it yield by voltaic action; its generating and consuming powers in relation to electric current are therefore equal. The amount of current produced by a given weight of a particular metal depends both upon the atomic weight and upon the degree of valency of the metal. An atomic weight of a monad metal yields one equivalent quantity of current; one of a dyad yields two; a triad three; and so on.

The percentage of equivalent of external current actually obtained is, however, in practice extremely variable, and the full proportion is rarely obtained. This arises from the circumstance that a greater or less proportion of the current generated circulates in minute local circuits upon the surface of the dissolving metal, and does not enter the external circuit at all. By actual experiment in nearly one hundred cases of various kinds, I found that the proportion of external current varied from about 2 to nearly 100 per cent.

Electrolytic Arrangements. — Various combinations and arrangements have been employed in which chemico-electric currents produce electrolysis; and these arrangements have been classified as follows :—1. Electrolysis by simple contact of one metal with one liquid; 2. By contact of one metal with two liquids; 3. By contact of two metals with one liquid; 4. By contact of two metals with two liquids; 5. By a separate electric current; and 6. By a separate current and a series of electrolysis vessels.

The first of these arrangements is termed the "simple immersion process," the most familiar example of which is the coating of iron with copper by simply dipping it into a solution of cupric sulphate. In this process the voltaic currents are excessively minute, are generated in immense numbers at points inconceivably small all over the immersed surface of the metal, and re enter producing electrolysis at all the intermediate points of that surface. In this arrangement the actions and products at the anodes cannot conveniently be observed or separated from those at the cathodes. The deposit of metal obtained by it is usually very thin.

The second consists in either carefully placing a lighter liquid in a distinct stratum upon a heavier one, or separating the two by means of a porous partition, and immersing the metal in contact with the two liquids. The portion of metal in one liquid then generates a current which re-enters the other part, or the second piece of the same metal in the second liquid, and produces electrolysis. By this contrivance the negative portion of the metal receives an electrolytic deposit in a liquid which the metal itself is unable to decompose by simple contact.

The third consists in bringing two metals into contact at their upper ends, either without or by means of a wire, and immersing their lower ends in the liquid; or allowing the metals to touch each other in the solution. Under these circumstances a current passes from the positive metal through the liquid into the negative one, producing electrolysis, and returns by the external circuit; the positive metal also acts simultaneously by "simple immersion process." This contrivance also enables the negative metal to receive an electrolytic deposit in a liquid which it does not decompose by "simple immersion," because the second metal offers a second path of return for the re-entering current. Cases of self-depositing metals acting by this process have long been recorded, in which a metal immersed in a solution of the same metal has produced a metallic deposit, *e.g.*, with cadmium in contact with copper in a boiling hot saturated solution of cadmic chloride the copper becomes coated with that metal. These cases have been but little investigated. Under this arrangement may be classed the "two metal" couples of Gladstone and Tribe, in which the resistance is greatly diminished, and therefore the strength of the current increased, by making the circuits indefinitely small. This is effected by electrolytically depositing copper, silver, or platinum in a porous spongy layer upon the surface of zinc or magnesium, washing the plate so prepared, and immersing it in the liquid to be electrolysed.

The fourth is termed the "single cell process," and consists of two liquids separated by a porous partition, the two metals being partly immersed, one in each liquid, and in contact with each other externally, or connected together outside by means of a wire. This method also enables a deposit to be produced upon a metal which does not decompose the liquid by simple contact. In this and the second arrangement, however, the liquids gradually diffuse into each other, waste the positive metal by simple immersion process, "and disturb the action at the negative surface."

The fifth is the most convenient arrangement, and the most frequently employed. It consists of a vessel containing the electrolyte and two electrodes, neither of which spontaneously decomposes the solution, the electrodes being connected with the battery or other source of current by means of two wires. It is known as the "battery process," or "separate current process." By it the strength of current in relation to the resistance in the electrolysis cell may be indefinitely increased, the most incorrodible metals may be used as anodes, and with a sufficiently dense current and suitable liquid even the alkali metals may be deposited. The sixth arrangement consists merely of a single series of such vessels and electrodes with an undivided current passing through the whole of them. It is not much employed.

Self-Deposition of Metals.—Raoult, also Gladstone and Tribe, have discovered some new cases of electrolysis of this kind. Raoult states (*Comptes Rendus*, Vol. LXXV., p. 1,103) that when two plates, one of copper and one of cadmium, are completely immersed in a solution of cadmic sulphate deprived of air, and covered with a layer of oil, as long as they do not touch each other, a very slight evolution of hydrogen is seen on the cadmium plate, whilst the copper shows no visible change. When, however, the plates are caused to touch each other, cadmium at once begins to be deposited on the copper one. Couples of gold iron, gold nickel, gold antimony, gold lead, gold copper, or gold silver, immersed either in cold or hot acid or neutral solution of salts of the more positive of these two metals, yielded no deposit of that metal (*Journal of Chemical Society*, 2nd series, Vol. XL, p. 464). Gladstone and Tribe also observed that a copper zinc couple separated zinc from a 1·5 per cent. aqueous solution of zinc sulphate (*ibid.* p. 453). Other instances of self-deposition will be given.

As these deposits of cadmium and zinc did not appear in solutions of the nitrates of those metals, and as an oxide of metal appears to be formed upon the corroded or positive plate, a probable explanation of the formation of the metallic deposits is that the water is decomposed, the salts in contact with the negative plate are reduced to the metallic state by the nascent hydrogen, and the acid thus formed is prevented from corroding the deposited metal by being immediately removed from it by diffusion into the mass of the liquid. Another arrangement in which a metal deposits itself is well known. It is that in which one metal is in contact with two different liquids, one of them being a solution of a salt of that metal.

Methods of Preparing Solutions for Electrolysis.—The exact details of preparing solutions for electro-chemical action differ of course in every different case. There are, however, two general methods—the one termed the chemical, and the other the battery or separate current process. In the former the usual processes of oxidation, crystallisation, solution, &c., are employed, and may be found sufficiently described in any work on general chemistry. The latter usually consists in taking a suitable solvent, hanging in it a large anode of the particular metal and a proper cathode, and passing a current until sufficient of the metal is dissolved and the liquid yields the desired deposit. The liquids obtained by the two processes, however, are not always exactly the same in chemical composition, because the electric process is attended by chemical changes at the cathode. In the latter process the anode is sometimes immersed in a portion of the liquid in a porous cell, the latter being partly immersed in the remainder of the solution.

Electro-Chemistry of Individual Substances.—Usually, electro-negative bodies appear at the anode, and electro-positive ones at the cathode; iodine, however, also sulphur, and less frequently selenium, appear either at the anode or cathode, according to the electric character of the body it is separated from, and in rare cases the same element is liberated at both electrodes simultaneously—*e.g.*, iodine from an aqueous solution of iodic acid.

The degrees of facility with which different substances are separated from their compounds, and the conditions of electrolytic balance of substances at their points of commencing separation, are subjects which have been but little examined, and require much investigation. Extensive tables will yet be formed showing the degrees of electromotive force in volts, and the density of current required to separate particular substances from certain liquids under given conditions. General truths will thus be evolved, throwing light upon the magnitude of the influence we term " chemical affinity " and upon the molecular relations of bodies, showing us also why some substances are easy and others difficult to separate. A systematic examination of the conditions under which allied substances, particularly chlorine and oxygen, are set free would probably enable us to determine those under which fluorine could be liberated. It is not by misdirected strength of current, however great, particular bodies are obtained, but by properly directed energy. From a weak solution of a potassium salt, even the strongest current with a large solid cathode will not secure to us the metal, but by using a cathode of mercury of small surface, even by means of a current of low electromotive force, potassium has been isolated. The laws of nature are universal, and electrolysis is no exception to the truth of the general statement that the chief secret of success in all things is well-directed energy.

As the study of electro-chemistry includes a knowledge not only of the conditions under which a given substance is electrolytically separated, but also of the electrolytic effect of a current or individual compounds, both are described in the following sections, and the series of substances are treated in systematic order.

Electrolytic Separation of Hydrogen.—H. Electro-chemical equivalent $= 1.00$ A monad cation. The only known gaseous metal. Is very readily separated; it is set free in a very large number of cases where water, acids, or other salts of hydrogen are electrolysed. In some cases it is set free by direct action, as when zinc or any other metal more electro-positive than hydrogen is immersed in the above liquids. In other cases it is liberated by "secondary action," as when those metals more electro-positive than hydrogen are electro-deposited from aqueous or acid solutions, and subsequently decompose the liquid by simple contact.

It is separated by all the electrolytic processes. Its liberation and subsequent spontaneous ignition when potassium is placed upon water is one of the most familiar and striking experiments of electro-chemistry. Magnesium, and especially its amalgam with mercury, decomposes water, setting free hydrogen. The same metal also liberates hydrogen from a great variety of saline solutions. Nearly all the readily oxidable metals decompose acidulated water, and the instances are so numerous that to specify all of them is quite unnecessary. According to H. St. Claire Deville, silver evolves hydrogen rapidly from aqueous hydriodic acid; and even silver, gold, and platinum, when in a finely divided state, liberate hydrogen from a hot concentrated solution of potassic cyanide.

I have observed the following cases relating to separation of the gas by magnesium. That metal did not evolve hydrogen by simple immersion in dilute hydrofluoric acid, and only a little in an aqueous solution of potassic chloride, but evolved it freely in a mixture of the two liquids. Similarly with the same acid and a solution of chlorate of potassium. It did not evolve the gas in a mixture of the same acid and a solution of potassic perchlorate. It set free hydrogen from a mixture of that acid and a solution of potassic bromide; but not from either alone. Similarly with magnesium in hydrofluoric acid mixed with solution of potassic iodide; but not with that acid when in admixture with solution of potassic iodate. It set free hydrogen from a mixture of that acid and solution of potassic sulphate; but not from either liquid alone. Probably the absence of gas was due, in some of these cases, either to the formation of a film of magnesic fluoride or suboxide upon the surface of the metal, and the insolubility of that salt in the particular liquid.

Some anhydrous hydrogen acids yield hydrogen readily by electrolysis; others do not. With platinum electrodes and a separate current from ten Smee cells, anhydrous hydrofluoric acid at 0°C. was freely decomposed; but anhydrous hydrochloric acid, liquefied by great pressure at 0°C., scarcely conducted at all, and evolved no visible gas.

Nearly all aqueous acids yield hydrogen at the cathode by the separate current process. This accords with G. Wiedemann's observation, that mixed liquids are more easily electrolysed than unmixed ones. According to Bourgoin, by electrolysis with platinum electrodes of distilled water containing pure sulphuric acid, the hydrogen and oxygen obtained are probably not results of an action of the current upon the water, nor of liberated electrolytic products acting upon the water, but of direct decomposition of a hydrate of sulphuric acid. Concentrated nitric acid does not liberate hydrogen by electrolysis, the hydrogen being absorbed.

Aqueous solutions of alkalis frequently yield hydrogen at the cathode. The electrolytic behaviour of each of the individual acids, salts, and alkalies, with regard to separation of hydrogen, &c., will be more fully described under the heading of the respective substances.

In consequence partly of the very frequent simultaneous deposition of hydrogen with other metals, those metals often contain that gas. It has been observed conspicuously in deposited palladium, and, to a less extent, in iron, cobalt, nickel, copper, and tin, and it has been stated that the explosive variety of deposited antimony contains hydrogen; but, according to E. Pfeifer, "explosive antimony" contains no free hydrogen (*Jour. Chem. Soc.*, Vol. XLII., 1882, p. 467). Much, however, depends, in all these cases, upon the kind of solution employed. I have several times observed that the steel blade of a knife which has been used as a cathode for a short time, either in a dilute acid, or in an alkaline liquid, becomes very brittle. Other investigators have also noticed that the simple immersion of iron in a dilute acid greatly reduces its tenacity; and it is not improbable that steam boilers are sometimes weakened by their decomposing the water and absorbing the hydrogen. It has been stated by Böttyer that if a piece of palladium, cobalt, nickel, or tin, has a wire of aluminium twisted round it, and is then immersed during a few minutes in a dilute acid, it absorbs sufficient hydrogen to exert a slightly reducing action upon a solution of potassic ferricyanide, also that a plate of palladium, previously coated with palladium black, absorbs the gas more rapidly, and when taken from the liquid and dried quickly between porous paper becomes red hot in the air in a few seconds.

For the absorption of hydrogen by platinum in electrolysis, see *Jour. Chem. Soc.*, 1877, Part II., p. 161, and for the deposition of hydrogen on both electrodes see *ibid.* Part I., p. 678.

Separation of Oxygen.—O. Electro-chemical equivalent $= \frac{16}{2} = 8$. A dyad anion. It is much less frequently or readily obtained than hydrogen by electrolysis, also less easily than the least oxidable metals. It is not separated by either of the electrolytic methods, except those in which a separate current is employed. To obtain it requires not only a separate source of current, but also an anode and liquid not easily oxidised. It is usually obtained by passing a current by means of platinum plates through a cooled mixture of one volume of sulphuric acid, and three to five volumes of pure water. The gas thus obtained is partly in the state of ozone, which may be detected by its odour. Numerous other mixtures of water with some acid, alkali, or salt, to render the mixture conducting, might be employed for obtaining it, but in all cases the anode must be a non-corrodible one. The electrolysis of various substances such as fused oxides, &c., which yield oxygen, will be described in their appropriate places.

Electrolysis of Water. — H_2O. Molecular weight = 18. F. Kohlrausch has shown (*Dingler's Polytechnik Journal*, Vol. 222, p. 283) that perfectly pure water is practically a non-conductor of the voltaic current (and probably not an electrolyte), and that on the addition of the least trace of impurity its conduction-resistance is greatly diminished.

Pure water is rapidly decomposed by simple immersion or contact of either of the alkali metals, less rapidly by aluminium amalgam and by magnesium, and slowly by the ordinary base metals, in each case by oxidation of the immersed metal. Magnesium amalgam containing one half a per cent. of magnesium decomposes water with violence, and more rapidly than sodium amalgam containing twice that percentage of sodium (Cailletet, Watts's "Dic. of Chem.," Vol. VI., p. 816). An amalgam of aluminium and mercury decomposes water at ordinary temperature (A. Cossa, Watts's "Dic. of Chem.," Vol. VII., p. 54). Alloys of aluminium and gallium decompose water readily, setting free much hydrogen and nearly the whole of the gallium as liquid metal (Lecoq de Boisbaudran, *Chem. News*, Vol. XXXVII., p. 274). Finely divided iron slowly decomposes boiling water, and sets free hydrogen (E. Ramann, *Jour. Chem. Soc.*, Vol. XL., 1881, p. 879). Water containing certain acids is decomposed more rapidly; boracic acid, cyanide of mercury, sugar, or gum dissolved in it have but little effect. The decomposition of water containing sulphuric acid, by means of zinc, is a common mode of obtaining hydrogen. Iron filings wetted with water, and exposed to the air or nitrogen at 60°F., induce the formation of ammonia (Berzelius).

Gladstone and Tribe state that pure water may be decomposed by a "copper-zinc couple," also by iron or lead which has been previously coated electrolytically with spongy copper. At 0°C. the decomposition of water by a zinc-copper couple is nearly *nil;* but at 100°C. it is very great (*Jour. Chem. Soc.*, Vol. XXXV., 1879, p. 572). With a magnesium platinum couple the decomposition is vigorous, even in cold water (*ibid.* p. 576).

According to D. Tommasi, water is not decomposed by a separate current from a single zinc-carbon or zinc-copper element if the electrodes are of platinum; but if the anode is a metal—copper, for instance—which, under the influence of that current, can unite with oxygen, the water is decomposed (*Jour. Chem. Soc.*, Vol. XLII., 1882, pp. 134 and 353).

The usual mode of electrolysing water is by previously mixing sulphuric acid freely with it, and passing a separate

current through the mixture by means of platinum elec-·
trodes; by this method its oxygen as well as its hydrogen is
obtained, and a small quantity of the former gas is absorbed
by the water. The oxygen contains a small proportion (but
not more than $\frac{1}{250}$ part of its weight) of ozone.

Janeczek considers that in the electrolysis of pure water,
hydrogen at the cathode and hydric peroxide at the anode are
the proximate resultants, and that the peroxide is resolved
into water and oxygen (*Jour. Chem. Soc.*, 1876, Part I., p. 182).

Bouvet electrolysed water under a pressure of several
hundred atmospheres. He found that the amount of water
decomposed by a given quantity of current was independent
of the pressure (*Jour. Chem. Soc.*, Vol. XXXVI., 1879, p. 293).

Water containing atmospheric air yields ammonia at the
cathode, and nitric acid at the anode (H. Davy).

Separation of Ozone.—Ozone is developed by electrolysis
in aqueous solutions of nitric, hydrofluoric, sulphuric, or phosphoric acids, also in those of nitre, potassic phosphate, or sodic
sulphate, but not in those of hydrochloric or hydrobromic
acids, or in strong nitric acid, or in aqueous solutions of
metallic chlorides, bromides, iodides, or ferrous sulphate.
According to Houzean (*Comptes Rendus*, Vol. LXXIV., p. 256),
the electrolysis of water furnishes only 3 to 5 milligrammes of
ozone per litre (*Jour. Chem. Soc.*, Vol. X., 2nd series, p. 220).

Electrolysis of Hydric Peroxide.—H_2O_2. Molecular weight
= 34. Electrolysis gradually resolves peroxide of hydrogen
into hydrogen and oxygen, the proportion of the latter being
greater than in the decomposition of water (Thenard).

E. Schone has electrolysed peroxide of hydrogen, and found
that the results were influenced by the strength of the solution, the degree of acidification, and the strength of the
current, and concludes that it is not an electrolyte, and that
its decomposition during electrolysis of the water or acid
present is a result of secondary action, due to the liberated
hydrogen and oxygen (*Jour. Chem. Soc.*, Vol. XXXVI., 1879,
p. 878).

According to Berthelot a dilute solution of hydric peroxide
undergoes electrolysis in two different ways—viz., one with,
and one without, the evolution of hydrogen, and both of
these may coexist. With high electromotive force both gases
are evolved, but with low electromotive force, such as that of
a zinc cadmium couple, only oxygen is given off, and no
hydrogen gas appears at the cathode. The latter decompo-
sitiou can be effected by any current, however feeble. In this
case either the peroxide splits up into water and oxygen, or
more probably a secondary action occurs, and the electrolytic
hydrogen combines with undecomposed peroxide to form
water.

Separation of Nitrogen.—N. Electro-chemical equivalent $\frac{14}{3} = 4\cdot66$. A triad anion. It is set free (along with other gases) by simple contact of metallic zinc with ammonic nitrate, in a state of fusion. A concentrated solution of ammonia, when electrolysed by a separate current and iron electrodes, yields pure nitrogen at the anode, and hydrogen at the cathode (Hisinger and Berzelius).

The electrolysis of compounds of nitrogen and hydrogen will be treated of with the alkali metals.

Electrolysis of Oxides of Nitrogen.—The only ones of these which appear to have been thus treated are hyponitric (N_2O_4) and ordinary nitric acid (HNO_3). The former aqueous acid conducts slowly, and is decomposed (Faraday). W. Zorn prepares hyponitrites by the electrolysis of a solution of a nitrite by means of a current from four Bunsen cells and mercury electrodes, and stopping the current as soon as ammonia begins to be evolved. In this reduction hydroxylamine is also formed (*Jour. Chem. Soc.*, Vol. XXXVIII, 1880, p. 4).

Nitric acid when concentrated is a good conductor. It yields with platinum electrodes oxygen, and simultaneously becomes yellow and then red at the cathode, and finally evolves gaseous nitric oxide. A more dilute acid yields hydrogen at the cathode, the quantity being greater as the acid is weaker and the current more dense; and if the acid is not of greater specific gravity than 1·24 and the current not too strong, the water alone of the acid is decomposed, and the full equivalent quantity of hydrogen is set free as gas.

By electrolysis concentrated nitric acid is decomposed with production of nitrous acid; with the acid of sp. gr. 1·2 a feeble current does not produce this effect. No ammonia is produced in dilute nitric acid, either *per se* or in presence of sulphuric acid; but if a solution of cupric sulphate is added in sufficient amount, sulphate of ammonium and metallic copper are simultaneously produced until all the nitric acid is converted into ammonium sulphate. In the presence of free alkali, nitrates are not converted into ammonia, but the latter is changed into nitric acid (C. Luckow, *Jour. Chem. Soc.*, Vol. XXXVIII., 1880, p. 282).

Brester states (*Chem. News*, Vol. XVIII., p. 144) that when decomposed by electrolysis nitric acid does not evolve any hydrogen gas at the surface of a cathode of platinum or charcoal; the acid is converted into ammonia. Bloxam (*Chem. News*, Vol. XIX., p. 289) has shown that the hydrogen set free from a cathode of platinum in dilute nitric acid, or in a solution of potassic nitrate, contained in a porous cell, placed in dilute sulphuric acid containing the anode, converts not more than one-half of the nitric acid of either of those solutions into ammonia. Bourgoin (*Comptes Rendus*, Vol. LXX., p. 811) has also electrolysed nitric acid.

The electrolysis of nitric acid, and solutions of its soluble salts with electrodes of wood charcoal or gas carbon yield mellogen free from nitrogen (Bartoli and Papasogli, *Jour. Chem. Soc.*, Vol. XLIV., 1883, p. 592; *The Electrician*, Vol. X., p. 388, Vol. XI., pp. 28 and 101).

In the electrolysis of red fuming nitric acid no gas is set free at first at either electrode. At the anode, NO_4 is totally changed to NO_5 by oxidation. At the cathode, NO_5 is reduced to H_3N during the whole of the electrolysis (A. Brester, *Chem. News*, Vol. XVIII., p. 145).

Finely divided copper, palladium, platinum, or carbon, charged with hydrogen, convert nitre into potassic nitrite and ammonia (Gladstone and Tribe, *Jour. Chem. Soc.*, Vol. XXXIII., 1878, pp. 306 and 307). Gladstone and Tribe have also investigated the electrolysis of a solution of potassic nitrate by a zinc copper couple, and are inclined to the hypothesis "that the two metals electrolyse the nitrate of potassium, with formation of nitrate of zinc, the reduction being effected at the negative pole through the agency of the potassium" (*Jour. Chem. Soc.*, Vol. XXXIII., 1878, p. 143). Professor Thorpe also has shown that the copper, zinc couple, in the presence of water and saltpetre, converts the whole of the nitrogen of the salt, first into nitrite and then into ammonia (*Jour. Chem. Soc.*, Vol. XXXIII., 1878, p. 139).

Passive State of Metals.—A peculiar condition, termed "the passive state," occurs with various metals when used as electrodes in nitric acid. By the following methods a platinum wire, to be used as the cathode in nitric acid of 1·49 sp. gr., and in which, with a suitable density of current, it would usually evolve gas for a time only, will be caused to evolve no gas from the moment of immersion. 1st. By connecting and then immersing the two polar platinum wires together in the liquid, and then at once separating them. (In this case, however, the acid must be diluted with less than its own volume of water.) 2nd. By igniting the cathode, and then immersing it after the anode. 3rd. By taking a second platinum wire, and after the cathode has ceased to evolve gas, joining the wire to it outside the liquid, then immersing the wire and withdrawing the cathode. The fresh cathode will then evolve no gas from the commencement, and this property may be transferred by it to a third wire, and a fourth one, and so on. A wire which has lost the power of liberating hydrogen recovers it by exposure to air, the time required being longer as the acid is stronger. In all these cases, if the current is too strong gas will be evolved. (Gmelin's "Handbook of

Chemistry," Vol. I., pp. 253-362. *See* also A. Brester, *Chem. News*, Vol. XVIII., p. 144.)

Separation of Fluorine.—F. Electro-chemical equivalent 19. A monad anion. I have made many attempts with this object by electrolysing anhydrous hydrofluoric acid, with anodes of carbon, platinum, palladium, and gold ; also by electrolysing certain fluorides in a state of fusion. In none of these cases, however, was that element definitely obtained. These experiments will be briefly described under the headings of the respective substances.

Electrolysis of Anhydrous Hydrofluoric Acid. — H.F. Molecular weight = 20. I have examined this highly dangerous and extremely volatile liquid. It boils at 67°F. Potassium immersed in the chilled acid evolved hydrogen, and produced vivid combustion. Sodium acted as it does upon water. The noble and base metals did not decompose it. Magnesium, aluminium, zinc, cadmium, tin, lead, reduced iron, powdered arsenic, antimony, or bismuth, did not expel hydrogen from it. I electrolysed the chilled fuming liquid by means of a separate current with a platinum anode; it conducted much more readily than pure water. With four Smee elements it began to conduct visibly, and with ten it conducted readily. No odour of ozone was evolved. The anode gradually acquired a thick red-brown crust, which deliquesced in the atmosphere. With forty elements the conduction was copious, the anode rapidly corroded, and much finely-divided platinum collected in the liquid. The brown coating was insoluble in the acid, but dissolved with formation of a basic salt in water, and formed a blood-red liquid. With an anode of very close-grained gas carbon, and six Smee cells, conduction occurred freely, and the carbon rapidly disintegrated. Anodes composed of fifteen different kinds of carbon of dense woods were tried with a current from ten elements; those made from kingwood, beech, ebony, boxwood, and lignum vitæ were the best. On immersing them in the acid, even without a current, they evolved bubbles (of air ?), cracked, and flew to pieces, and on passing a current they broke immediately, some with violence, projecting the fragments and liquid in all directions —even the densest kinds behaved thus. The most resisting was that made from beechwood. With much difficulty, and by the aid of a magnesium light, it was ascertained that the passage of the current was not attended by any increase of bubbles from the carbon. No special odour besides that of the acid could be detected, but the charcoal, when removed from the liquid, emitted a feeble chlorous odour, as well as that of the acid.

With forty Smee elements and an anode of gold the acid
scarcely conducted at all ; in half an hour the gold was somewhat corroded, and acquired on its edges a few green crystals,
which became red by contact with the moisture of the atmosphere. With a palladium anode the acid conducted more
freely, but less so than with one of platinum or charcoal. A
current from forty Smee elements caused a palladium anode
to corrode, and become covered with a thick brittle crust of a
dark red-brown colour upon its outer surface and a brighter
red beneath. By prolonged action a quantity of this substance
was collected on a plate of platinum upon a heated block of
iron, and was subsequently investigated.

In each of these experiments the acid was contained in a
large platinum cup immersed in a freezing mixture. The cup
was provided with a lid of paraffin to exclude moisture, for
which the acid has most intense attraction ; it was also divided
vertically in the middle by a plate of paraffin, which extended
to within about half an inch of the bottom of the vessel, in
order to prevent evolved hydrogen touching the anode deposit
and rapidly reducing it to metal.

Electrolysis of Aqueous Hydrofluoric Acid.—According to
Faraday, aqueous hydrofluoric acid is not decomposed by
electrolysis, but only the water in it. I electrolysed the pure
dilute liquid containing about 10 per cent. of the acid, by
means of a separate current and sheet platinum electrodes.
Gas was evolved freely from each electrode, and a very strong
odour of ozone was observed. No corrosion of either electrode
occurred during twelve hours' action. The gas from the anode
was collected ; it re-inflamed a red hot splint vividly ; paper
wetted with spirits of turpentine was not blackened, nor was
bright silver tarnished by it; it was oxygen. I similarly
electrolysed, by a current from ten Smee cells, the pure
aqueous acid containing about 80 per cent. of the anhydrous
substance. Copious conduction took place, with much evolution of oxygen at the anode. Heat was produced in the liquid ;
the anode dissolved slowly ; in three hours it lost 1·58 grain.
The smell of ozone disappeared if the electric current was
much weakened, and reappeared on first contact. In a
further eleven hours, the anode lost 5·05 grains, and was
covered with a blackish crust which was partly soluble in
water to a brownish solution. In a further twenty hours the
loss had increased from 5·05 to 15·00 grains, without any signs
of metallic deposit upon the cathode.

I also electrolysed during five hours a chilled mixture of
160 grains of the anhydrous acid, 244·4 grains of concentrated nitric acid, and 273·8 grains of pure water, by means
of sheet platinum electrodes and six Smee elements. Free
conduction occurred, and much odourless oxygen was evolved.

The anode was not corroded, and no gas was visible at the cathode. By similar electrolysis with a current from ten Smee cells of a mixture of equal volumes of 30 per cent. pure aqueous hydrofluoric acid and strong hydrochloric acid, much chlorine was set free from the anode and hydrogen from the cathode. This is consistent with the usual effect that chlorides, like oxides, are decomposed before fluorides. A mixture of equal volumes of the aqueous acid and strong oil of vitriol yielded much oxygen and a strong odour of ozone at the anode and hydrogen freely at the cathode. The anode corroded very slowly, and fumes were evolved which rapidly blackened gutta percha. With selenious acid in place of the sulphuric, gas was set free at both electrodes, and much red selenium was deposited upon the cathode. No odour of ozone was evolved until a large quantity of red and black selenium had been deposited; it was then evolved freely. The anode was not corroded during twenty-eight hours' free electrolysis. By electrolysis of the dilute hydrofluoric acid, to which some phosphoric anhydride had been added, ozone was evolved from the anode and hydrogen from the cathode; the anode was also slowly corroded.

Bartoli and Papasogli have also electrolysed aqueous hydrofluoric acid with anodes of wood charcoal or gas carbon, and found the anodes disintegrate (*The Electrician*, Vol. XL, pp. 28 and 101; *Jour. Chem. Soc.*, Vol. XLIV., 1883, p. 590).

Separation of Chlorine.—Cl. Electro-chemical equivalent = 35·5. A monad anion. Set free on passing, by means of an anode of carbon or platinum, an electric current through concentrated hydrochloric acid, or through aqueous solutions of the chlorides of sodium, ammonium, or other metals, also through various chlorides in a state of fusion. With aqueous solutions, some of the chlorine usually dissolves in the liquid. The electrolysis of chlorine water yields hydrochloric acid at the cathode and a little chloric acid at the anode (Balard, also Connell, Gmelin's "Handbook of Chemistry," Vol. I., p. 451).

Electrolysis of Hydrochloric Acid.—HCl. Molecular weight = 36·5. I ascertained by experiment (*Proc. Roy. Soc.*, May 4, 1865) that the anhydrous substance, liquefied by great pressure, is a very feeble conductor of electricity. Two fine platinum wires immersed in it $\frac{3}{8}$ths of an inch in length and $\frac{1}{10}$th of an inch asunder, and connected with ten Smee elements, evolved no perceptible bubbles of gas, and produced only a small deflection amounting to 23° of the needles of a sensitive galvanometer; and this amount of conductivity might possibly have been due to a minute trace of oil of vitriol mixed with the liquid acid. In a second similar experiment, with the wires $\frac{1}{16}$th of an inch apart, not the slightest conduction occurred on using the same battery power, but by employing the secondary current of a strong induction coil, with condenser attached, conduction and a steady deflection of 20° of the needles took place, gas being freely evolved from the negative wire only. It is evident, therefore, that liquefied hydrochloric acid is a very bad conductor of electricity. Bleekrode subsequently discovered (*Proc. Roy. Soc.*, Vol. XXV., 1876, p. 325) that the anhydrous liquefied acid "opposes a formidable resistance, and is not decomposed in a perceptible way" by the passage through it of a current from 5,640 cells of De la Rue's chloride of silver battery.

Gallium liberates hydrogen freely by simple immersion in dilute hydrochloric acid (M. Lecocq de Boisbaudran).

Electrolysis of concentrated hydrochloric acid with a platinum anode causes the anode to dissolve, but that of the dilute acid causes the formation of chlorine compounds at the anode without corroding the platinum (D. Tommasi, *Jour. Chem. Soc.*, Vol. XLIV., 1883, p. 142).

In dilute solutions of metallic chlorides by electrolysis hypochlorous acid is alone produced; in concentrated ones chlorine is also set free. Chlorates are produced from the chlorides of the alkalies and alkaline earths, as soon as the reaction of the solutions has become alkaline, from the evolution of the chlorine and hypochlorous acid (C. Luckow, *Jour. Chem. Soc.*, Vol. XXXVIII., 1880, p. 282). If dilute chloride solutions contain a little free hydrochloric acid, hypochlorous acid is alone produced, and the solution, after a time, acquires an alkaline reaction.

The electrolysis of aqueous solutions of certain metallic chlorides by means of the contact of two metals has been investigated by Gladstone and Tribe (*Phil. Mag.* [4], Vol. XLIX., p. 425), and will be described under the headings of the respective metals. Thorpe has shown that the copper zinc couple reduces chlorate of potassium to chloride (Gladstone and Tribe; *Jour. Chem. Soc.*, Vol. XXXIII., 1878, p. 147). Platinum charged with hydrogen behaves similarly (*ibid.*, p. 309), but more powerfully.

Electrolysis of Oxides of Chlorine.—Very little has been done in this part of the subject. Aqueous solution of oxide of chlorine (ClO_2) yields hydrogen at the cathode and a small quantity of oxygen gas and perchloric acid at the anode. (Count Stadion). I have electrolysed aqueous chloric and perchloric acids with anodes of silver.

Separation of Bromine.—Br. Electro-chemical equivalent $= 80$. A monad anion. It is separated in many cases when aqueous solutions of bromides are electrolysed by means of a separate current and an incorrodible anode. A portion of the liberated bromine usually dissolves in the liquid, an aqueous solution of bromine yields by electrolysis hydrobromic acid, and a mere trace of hydrogen at the cathode, but no bromic acid at the anode. The water is decomposed (Balard, also Connell, Gmelin's "Handbook of Chemistry," Vol. I., p. 451).

Electrolysis of Hydrobromic Acid.—HBr. Molecular weight = 81. Bleekrode has stated (*Proc. Roy. Soc.*, Vol. XXV., p. 323) that anhydrous hydrobromic acid is a non-conductor to the voltaic current from eighty Bunsen elements. The aqueous acid when electrolysed by a separate current liberates bromine at the anode and hydrogen at the cathode.

Electrolysis of Oxides of Bromine.—I have been unable to find any record of any one having electrolysed either bromic or perbromic acids, or aqueous solutions of their salts. By immersing a sheet of aluminium in an aqueous solution of bromic acid, I observed that hydrogen and bromine were set free.

Separation of Iodine.—I. Electro-chemical equivalent = 127. A monad anion. It is, however, sometimes separated by secondary action at the cathode. According to Bleekrode (*ibid.*) liquid anhydrous hydriodic acid does not transmit any current from eighty Bunsen elements. Faraday observed that by electrolysis, with a separate current, of potassic iodide, or iodide of lead, in a state of fusion, iodine was set free at the anode. A solution of iodine in water yields by electrolysis some hydrogen at the cathode. The water is decomposed (Balard, also Connell, Gmelin's "Handbook of Chemistry," Vol. I., p. 451).

Electrolysis of Aqueous Hydriodic Acid.—A concentrated solution of aqueous hydriodic acid yields by a separate current, and platinum electrodes, iodine alone at the anode; but a dilute one yields iodine and oxygen (Faraday). Matteucci observed that the stronger the current, and the more dilute the acid, the greater was the proportion of oxygen.

If the solution of an iodide be covered with starch jelly, the cathode be placed in the former, and the anode in the latter, the starch is turned blue around the anode, even if the solution contain a much larger quantity of bromide or chloride than of iodide (Steinberg, *Jour. Pr. Chem.*, Vol. XXV., p. 288; Watts's "Dictionary of Chemistry," Vol. III., p. 287).

Riche states (*Comptes Rendus*, Vol. XLVI., p. 348) that iodic acid (HIO_3) is produced by electrolysis of aqueous iodine, or an aqueous solution of hydriodic acid. In the latter case the acid is simply oxidised to iodic acid by oxygen evolved by the decomposition of water. In the former case the iodine is first converted into hydriodic acid, and then oxidised in this way.

Electrolysis of Oxides of Iodine, &c.—By immersing a sheet of aluminium in a solution composed of twenty-six grains of dry iodic acid and five ounces of water, I observed that much gas was evolved; the metal acquired a strong odour of absorbed iodine, and had increased about 16 per cent. in weight (*Proc. Birm. Phil. Soc.*, Vol. IV., Part I.)

A solution of one part of iodic acid and ten parts of water yields oxygen gas at the anode and iodine alone at the cathode, the latter being separated by secondary action of hydrogen, liberated by the electrolysis of the water (Connell). According to Buff, however (*Ann. Chem. et Pharm.*, Vol. CX., p. 257), the iodic acid is resolved by the current into hydrogen and iodic anhydride, which latter is decomposed by the water, thus producing iodic acid and free oxygen (Watts's "Dictionary of Chemistry," Vol. III., p. 300). The electrolysis of periodic acid does not appear to have been yet examined.

When an iron or copper plate, or better, a zinc and copper plate, connected externally by a wire, are immersed in strong solution of potassium iodate at 60° F., complete reduction to potassic iodide occurs. Potassium bromate is similarly reduced to bromide; but potassic chlorate slowly and incompletely to chloride (G. Pellagri, Watts's "Dictionary of Chemistry," Vol. VIII., Part 2, p. 1,668).

Electrolysis of Bromide of Iodine.— When an aqueous solution of starch and iodine, which has been turned yellow by dissolved bromine, is subjected to electrolysis, it becomes orange coloured at the anode by liberation of bromine, and blue at the cathode by separation of iodine (De la Rive, *Ann. Chem. et Phys.*, Vol. XXXV., p. 164).

Electrolysis of Iodides, Bromides, and Chlorides.—By electrolysis, iodine and bromine are separated from solutions of iodides and bromides. Iodates and bromates are produced simultaneously from the iodides and bromides of the metals of the two first groups, especially in concentrated solutions. When the solutions of the chlorides, bromides, and iodides contain free alkali, only chlorates, bromates, and iodates are produced. From the insoluble compounds of chlorine, bromine, and iodine, with the metals, suspended in dilute sulphuric or nitric acid, the acid radicle appears at the anode and the metal at the cathode (C. Luckow, *Jour. Chem. Soc.*, Vol. XXXVIII., 1880, p. 282).

Separation of Carbon.—C. Atomic weight = 12. A tetrad cation. "An excess of silicon fused with potassic carbonate sets free carbon." Deville states that metallic aluminium liberates carbon from carbonate of potassium in a state of fusion (*Chemist,* New Series, Vol. IV., p. 481). I have observed the same with fused sodic carbonate. According to Phipson, magnesium by contact with fused carbonate of sodium set free carbon abundantly (*Proc. Roy. Soc.*, 1864, Vol. XIIL, p. 217, also *Chemical News*, Vol. IX., p. 219).

The following are some experiments of mine (*Proc. Birm. Phil. Soc.*, Vol. IV.) :—A fused mixture of 200 grains of pure sodic hydrate, 170 grains of pure precipitated silica, and 610 grains of the mixed anhydrous carbonates of sodium and potassium was electrolysed by means of a current from ten Smee elements, with a sheet platinum anode and a thick platinum wire cathode. Conduction was free, and much oxygen, which relighted a red-hot splint, was liberated at the anode. Dark streams flowed from the cathode, sodium was also set free, and if the cathode was only slightly immersed bubbles of vapour of sodium were emitted, and took fire at the surface of the liquid. After one hour's action the platinum anode had lost ·37 grain in weight. The cathode had a feebly adherent rough deposit of a dull jet black colour upon it. This deposit was subsequently washed and dried ; a portion of it burned with a glow when heated to redness, and left a minute residue of grey platinum; it also deflagrated with fused nitre below a red heat, and vividly by heating with potassic chlorate. It did not dissolve nor evolve any gas in a mixture of strong nitric acid and pure concentrated hydrofluoric acid. It was, therefore, carbon.

As carbon was not readily deposited from the fused carbonates of potassium and sodium, whilst silicon was deposited from fused silicofluoride of potassium, and as " an excess of silicon fused with potassic carbonate sets free carbon, but silicon with an excess of the carbonate liberates carbonic oxide," the carbon liberated in this experiment may have been a secondary result, and an effect of previously deposited silicon reacting upon the fused mixture. It was with the expectation of this effect that I employed silica in the mixture.

I also electrolysed in a platinum cup a fused mixture of 475·2 grains of 97·1 per cent. sodic carbonate (containing as impurity only water) and 217·4 grains of borofluoride of sodium, by means of the same current, a sheet platinum anode and thick platinum wire cathode. Conduction was free. Gas arose from the anode, and a small amount of black deposit formed upon the cathode. After having been well washed the deposit was dried, put on a platinum dish, and heated to redness ; it burned with sudden incandescence until nearly the whole was consumed. It was, therefore, nearly wholly carbon.

I electrolysed in a platinum cup a fused mixture of 274 grains of pure sodic carbonate, 375 grains of pure potassic carbonate, both anhydrous, and 206 grains from crystallised boracic acid, at a red heat, by means of a current from eight Smee elements and platinum electrodes. There was free conduction, much gas from the anode, and an instant jet black deposit formed upon the cathode, and could be burned off at a red heat. Metallic sodium was set free at the cathode, especially during deep immersion. The anode was soon much corroded, and acquired a very smooth surface, and platinum was deposited upon the cathode. No free carbon was ultimately found.

Electrolysis of Carbonic Anhydride.—CO_2. Molecular weight = 44. I have examined (*Phil. Trans. Roy. Soc.*, 1861) the action of a voltaic current on carbonic anhydride liquefied by great pressure. With electrodes of thin platinum wire $\frac{1}{10}$th of an inch apart, and the liquid below 32°F., not the slightest conduction occurred with a current from forty Smee elements; and sparks from a Ruhmkorff coil, which passed through $\frac{3}{8}$ths of an inch of cold air in an alternate portion of the divided circuit, would not pass through the liquid. In another trial, with the wires about $\frac{1}{70}$th of an inch asunder, sparks from the coil, which were passing freely through $\frac{9}{32}$nds of an inch of cold air in the alternate circuit, passed occasionally through the cold acid and exhibited a pale blue colour. The liquid is, therefore, a strong insulator of electricity. Bleekrode also (*Proc. Roy. Soc.*, 1876, p. 325) tried the same liquid with a current from 5,540 chloride of silver elements. A spark jumped between the poles, and the tube exploded. He concluded that the liquid is a very bad conductor. Cailletet (*Comptes Rendus*, Vol. LXXV., p. 1,271) has also arrived by experiments at the same conclusion.

I tested by experiment in an approximate manner the relative degrees of conduction resistance of distilled water, and of the same saturated with carbonic anhydride at 60°F. and at atmospheric pressure. No conspicuous difference was observable.

I also passed an electric current from four Smee elements by means of platinum wires during one week through very dilute sulphuric acid in a large U glass tube, one leg of which was kept full of a mixture of carbonic oxide and carbonic anhydride gases. No carbon was deposited. Fuming sulphuric acid, also a syrupy solution of phosphoric acid, were saturated with dry carbonic anhydride, and then electrolysed by means of platinum wire electrodes and currents from 112 Smee cells in single series; no carbon was deposited (*Proc. Birm. Phil. Soc.*, Vol. IV.).

Electrolysis with Anodes of Carbon.— According to A. Bartoli and G. Papasogli, in liquids whose electrolysis is not accompanied by evolution of oxygen at the anode, anodes of wood charcoal, gas carbon, and graphite are not disintegrated or dissolved, or suffer any loss of weight. In those in which oxygen is evolved, those anodes are partly disintegrated, and partly oxidised to carbonic oxide and carbonic acid gases, together with other products; graphite used in those liquids never imparts a colour to the electrolyte, but anodes of wood charcoal and gas carbon, previously purified, colour it black, both in alkaline solutions and in those of certain acids and salts, by the formation of a black substance which they term mellogen, the composition of which is represented by the formula $C_{11}H_2O_4$, together with traces of benzo-carboxylic acid; graphite anodes in those liquids produce graphitic acid $C_{14}H_2O_3$. In alkaline electrolytes, anodes of wood charcoal, gas carbon, and graphite produce mellic acid $C_{12}H_6O_{12}$; pyro-mellic acid $C_{10}H_6O_8$; hydromellic acid $C_{12}H_{12}O_{12}$, and another body, apparently hydro-pyromellic acid, $C_{10}H_{10}O_8$ (*Jour. Chem. Soc.*, Vol. XLIV., 1883, p. 592; *The Electrician*, Vol. XI., pp. 28 and 101). For the electrolysis with electrodes of wood charcoal, gas carbon, and graphite, of solutions of hydrochloric, hydrobromic, and hydriodic acids and their potassium salts, potassic cyanide, sulphuric and nitric acids and their salts, hydrogen and sodium sulphite, arsenic acid, boracic acid, alkaline hypochlorites, permanganates, bichromates, and chlorates, chromic acid, mellic acid, oxalates, formiates, acetates, &c., and sodic pyrogallate, *see* also the same paper.

Separation of Boron.—B. Atomic weight = 31, equiv. $\dfrac{10\cdot9}{3}$

= 3·63. A triad cation. By contact of magnesium with boracic acid in a fused state boron is set free (Phipson, *Proc. Roy. Soc.*, Vol. XIII., 1864, p. 217; also *Chem. News*, Vol. IX., p. 219).

"Boron was first electro-chemically isolated by Sir H. Davy. He states that when boracic acid is exposed between two surfaces of platinum, receiving at the same time all the action of a current from 300 cells, an olive brown matter is formed upon the negative surface, gradually increasing in thickness, and finally becoming black. The isolated body is boron" (*Chem. News*, Vol. XII., p. 3).

Electrolysis of Oxide and Fluoride of Boron.—Burckhard states (*Chem. News*, Vol. XXI., p. 238) that pure boracic acid in a state of fusion is a non-conductor. I found that by electrolysing pure borofluoride of potassium in a fused state, with platinum electrodes and a separate current, boron was deposited, and combined with the cathode, rendering the latter rough and brittle.

Separation of Silicon.—Si. Atomic weight = 28. A tetrad cation. According to Golding Bird (*Phil. Trans. Roy. Soc.*, 1837, p. 37) silicon may be electro-deposited from a solution of its fluoride in alcohol. The kind of apparatus he employed was a combination of one voltaic cell in undivided circuit, with a "single cell apparatus," the silicon being deposited upon the negative platinum plate of the latter. I electrolysed in a platinum cup a fused mixture of 300 grains of 97·1 per cent. pure potassic carbonate (the 2·9 per cent. being water) and 442 grains of silico fluoride of potassium, by a current from ten Smee cells, a sheet platinum anode, and a platinum wire cathode. Gas arose from the cathode at first only ; after that streams of black matter poured down from the cathode, and the latter acquired a blackish film, but subsequently alloyed with silicon, and fused on its surface.

Separation of Hydride of Silicon.—This compound is obtained, in admixture with much free hydrogen, when the current from 8 to 12 Bunsen elements is passed by means of an anode of aluminium containing silicon into an aqueous solution of common salt. The aluminium dissolves as chloride, setting free much gas, some of the bubbles of which inflame spontaneously in the air, emitting a white light, and diffusing finely divided silica. The compound appears to be due to a secondary action ; a part of the nascent hydrogen, set free by union of aluminium with the oxygen of the water, unites with the silicon (Wöhler and Buff, *Ann. Chem. et Pharm.*, Vols. CII., CIII., CIV., and CXII.). It is not stated whether this compound would be formed by simple immersion without a separate current. I have observed that a lump of fused silicon, immersed in a mixture of pure hydrofluoric acid (strong) and nitric acid, evolves a spontaneously inflammable gas.

By electrolysis, silicic and boric anhydrides are separated from their concentrated solutions at the anode (C. Luckow, *Jour. Chem. Soc.*, Vol. XXXVIII., 1880, p. 283).

Becquerel has investigated the decomposition of silicates and other minerals by electro-capillary diffusion (*Comptes Rendus*, Vol. LXVII., p. 1,081).

Separation of Titanium.—Ti. Atomic weight = 50. A cation. This element does not appear to have yet been electro-deposited, nor its compounds electrolysed. I observed that crystals of nitro-cyanide of titanium conducted freely a current from 60 Smee elements.

For the electrolytic analysis of zirconium, *see* A. Claessen, *Jour. Chem. Soc.*, Vol. XLII., 1882, p. 896.

Separation of Sulphur.—S. Atomic weight = 32. A dyad anion ; sometimes also separated by secondary action at the cathode. Obtained by electrolysing, by means of a separate current and platinum electrodes, an aqueous solution of sulphide of potassium, hydrogen being simultaneously set free at the cathode (Faraday).

MM. Blas and Miest have shown that if in electrolysis we replace the anodes of metal by metallic sulphide ores compressed to hard plates, and use a suitable electrolyte, all the sulphur of the ore is separated at the anode and falls down, and the metal is deposited upon the cathode (*Chem. News*, Vol. XLVI., pp. 93 and 121 ; *The Electrician*, Vol. X., p. 388).

Hydric Sulphide.—H_2S. Molecular weight = 34. This substance when liquefied by great pressure does not appear to have been yet subjected to the action of an electric current. Its aqueous solution would no doubt yield sulphur at a platinum anode. When dilute sulphuric acid is electrolysed with a zinc anode and a charcoal cathode, hydric sulphide is evolved at the latter (Highton, *Chem. News*, Vol. XXVI., p. 117).

Electrolysis of Sulphur Dioxide.—SO_2. Molecular weight = 64. This oxide, liquefied by pressure, does not transmit a current from 40 cells. An aqueous solution of the gas yields sulphur and hydrogen at the cathode by the passage of such a current (De la Rive, Gmelin's " Handbook of Chemistry," Vol. II., p. 170).

The electrolysis of its aqueous solution, H_2SO_3, is not simply a separation of the oxide into oxygen at the anode and sulphur at the cathode. According to A. Guerout (*Comptes Rendus*, Vol. LXXXV., p. 225), with a feeble current H_2SO_4 is produced at the anode and a yellow liquid at the cathode ; with a stronger one sulphur also appears with the yellow liquid, and with a still stronger sulphur alone is deposited at the cathode. Its electrolysis resembles that of a salt, the acid and oxygen being set free at the anode, and the hydrogen (H_2) appearing at the cathode, where it acts upon a fresh portion of the acid, and reduces it thus :—$H_2 + H_2SO_3 = H_2SO_2 + H_2O$. This agrees with the fact that hyposulphurous acid (H^2O_2S) and sulphur appear at the cathode, the sulphur being produced by the decomposition of that acid formed there in a concentrated state (*Chem. News*, Vol. XXXVI., p. 90).

Sulphurous acid in aqueous solution is decomposed by the current into sulphur and sulphide of hydrogen, and sulphites are gradually converted into sulphates. Thiosulphates are converted into their corresponding sulphates with separation of sulphur. The alkaline sulphides, according to their richness in sulphur, are decomposed with or without separation of sulphur, sulphates being formed. In the alkaline sulphates and thiosulphates, in addition to sulphides, polythionates are always produced (C. Luckow, *Jour. Chem. Soc.*, Vol. XXXVIII., 1880, p. 283).

A copper zinc couple liberates sulphur from sulphurous acid without producing sulphuretted hydrogen (Gladstone and Tribe, *Jour. Chem. Soc.*, Vol. XXXIII., 1878, p. 307).

Electrolysis of Sulphuric Acid. — H_2SO_4. Molecular weight = 98. Sulphuric anhydride (SO_3) is a non-conductor with a current from 14 Bunsen cells. Its solution in concentrated oil of vitriol is decomposed by a separate current with platinum electrodes into oxygen at the anode and sulphur at the cathode. By varying the proportion of the two substances, part of the sulphur reduces the sulphuric acid to sulphurous anhydride, which is evolved at the cathode (Geuther, *Ann. Chem. et Pharm.*, Vol. CIX., p. 130).

By electrolysis, concentrated English sulphuric acid is decomposed with deposition of sulphur (C. Luckow, *Jour. Chem. Soc.*, Vol. XXXVIII., 1880, p. 283).

Separation of Persulphuric Acid.—S_2O_7. Berthelot divided two portions of diluted and chilled sulphuric acid by a porous partition, immersed stout platinum wire electrodes in the two portions, and passed a dense current from three very large Bunsen cells through the liquids, and thus obtained a mixture of dilute sulphuric acid containing 88 to 123 grammes of S_2O_7 per litre (*Comptes Rendus*, No. VII., February 16th, 1880; *Jour. Chem. Soc.*, Vol. XXXVIII., 1880, p. 607).

Liquid chloride of sulphur, and also carbonic bisulphide, are non-conductors.

Separation of Selenium.—Se. Atomic weight $= 79.5$. A cation; acts also as an anion. Very little investigation has yet been made of the electrolysis of compounds of this element. A mixture of aqueous hydrofluoric and selenic acids yielded much red selenium upon the cathode. During the electrolysis by a separate current of an aqueous solution of selenate of nickel, containing selenate of sodium and free selenic acid, I repeatedly observed an abundant deposit of bright red selenium upon a platinum cathode. The deposition was no doubt due to decomposition of the free acid, because it ceased on neutralising the acid with ammonia. According to L. Schicht (*Chemisches Centralblatt*, No. XXIV., 1880; also *Berg und Hüttenmannische Zeitung*, 1880), selenium is readily and completely reduced and thrown down by a feeble current from not more than two cells, both from acid and alkaline solutions (*Chem. News*, Vol. XLI., p. 280, Vol. XLII., p. 331, and *English Mechanic*, Vol. XXXI., p. 540).

Separation of Tellurium.—Te. Atomic weight $= 129$. A triad cation. Ritter, and subsequently Sir H. Davy, observed whilst electrolysing water with a tellurium cathode that the water around the cathode acquired a purple colour by dissolving telluride of hydrogen, and then precipitated a brown powder. Magnus showed that the brown powder was metallic tellurium set free by oxygen, which diffused from the anode, and decomposed the telluride. If the water is acid the telluride does not dissolve, but escapes as gas.

L. Schicht states (*ibid.*) that tellurium is readily and completely thrown down both from acid and alkaline solutions, but more readily than selenium. From an acid solution it is easily deposited with a blue-black colour, and from alkaline ones it is separated in a very loose state at the anode with much evolution of gas, and if much metal is present it floats as a light powder upon the liquid.

Electrolysis of Telluric Fluoride and Chloride.—I have electrolysed pure dilute hydrofluoric acid with an anode of pure tellurium and a current from a single Smee element. The action was very slow, and most excellent deposits of bright reguline metal of grey colour and bright crystalline structure were obtained. By electrolysing a pure solution of telluric chloride by means of a very feeble current and large electrodes of smooth platinum I obtained only a jet black deposit, chiefly of non-adherent metal.

For the electrolytic purification of tellurium, *see* Watts's "Dictionary of Chemistry," Vol. VIII., Part 2, p. 1,895.

Separation of Phosphorus.—P. Atomic weight = 31. A triad element. Acts both as an anion and a cation. According to Burckhard (*Chem. News*, Vol. XXL, p. 238), fused pyrophosphate of sodium yields by electrolysis with a separate current phosphorus and oxygen at the anode and soda at the cathode; if the anode is composed of platinum a phosphide of that metal is formed.

Electrolysis of Oxides of Phosphorus.—The electrolysis of concentrated phosphoric acid produces a metallic phosphide with the cathode when the latter is composed of copper or platinum (H. Davy).

By electrolysis with platinum electrodes dilute solutions of phosphoric acid or phosphates undergo no change (C. Luckow, *Jour. Chem. Soc.*, Vol. XXXVIII., 1880, p. 283).

The electrolysis of phosphoric acid and solutions of its salts, with electrodes of wood charcoal or retort carbon, produces phospho-mellogen, and with graphite electrodes phosphographitic acid (Bartoli and Papasogli, *Jour. Chem. Soc*, Vol. XLIV., 1883, p. 592; *The Electrician*, Vol. XL, pp. 28 and 101).

Chlorides, Bromides, and Iodides of Phosphorus.—These are non-conductors of a voltaic current.

Separation of Arsenic.—As. Atomic weight = 75. A triad cation. Easily separated by various electrolytic processes. Palladium charged with hydrogen reduces a solution of arsenious acid to metal without producing arsenide of hydrogen (Gladstone and Tribe, *Jour. Chem. Soc.*, Vol. XXXIII., 1878, p. 308). It is also separated—1. By dissolving arsenious acid in warm dilute hydrochloric acid and stirring the solution with a piece of clean copper the latter acquires a coating of arsenic; this is the well known "Reinsch's test" for the element; 2. By contact of zinc with platinum in solutions of arsenic the latter is deposited upon the platinum; and 3. By passing a separate current through a solution of arsenic in dilute hydrofluoric acid, by means of an anode of arsenic and a cathode of platinum, I have obtained a scaly deposit of the metal.

The electrolysis of arsenic acid and solutions of its soluble salts and electrodes of wood charcoal, or gas carbon, yields mellogen free from arsenic (Bartoli and Papasogli, *Jour. Chem. Soc.*, Vol. XLIV., 1883, p. 592; *The Electrician*, Vol. XL, pp. 28 and 101).

Separation of Arsenide of Hydrogen.—AsH_3. Molecular weight = 78. From acid solutions of arsenic, magnesium by simple immersion evolves this poisonous and inflammable compound (Roussin, *Chem. News.* Vol. XIV., p. 27). Marsh's test for arsenic consists in evolving this gas from an acid solution of arsenic by simple immersion of zinc in it. The same gas is evolved at a platinum cathode by the passage of a separate current through solutions of arsenic, when the current is sufficiently strong. A solid hydride of arsenic, supposed to have the composition $As.H_2$, is produced when water is electrolysed by a strong current, with metallic arsenic for the cathode Watts's "Dictionary of Chemistry," Vol. III., p. 181).

Terchloride of Arsenic.—$AsCl_3$. Molecular weight = 181·5. This liquid is a non-conductor of a voltaic current, but the aqueous solution conducts readily and is decomposed.

For the electrolytic analysis of arsenic, see *Jour. Chem. Soc.*, Vol. XLII., 1882, p. 1,320; also *Chem. News*, Vol. XLVI., p. 106. And for the detection of arsenic in mineral waters by means of a voltaic couple of tin and gold, see J. Lefort, *Jour. Chem. Soc.*, Vol. XXXVIII., 1880, p. 510.

Separation of Antimony.—Sb. Elec. chem. eqt. $\frac{120}{3} = 40 \cdot 00$.

A triad cation. This metal may be obtained from its solutions by all the methods of electrolysis. It is easily deposited from an acid solution of its terchloride by simple contact of various metals. Zinc, bismuth, tin, lead, brass, and German silver were coated with antimony by simple immersion in that solution; but platinum, gold, silver, nickel, and antimony were not. The simple immersion process is used to impart a lilac colour to articles of brass. A small quantity of hydrochloric acid, which has been perfectly saturated with freshly precipitated and wet teroxide of antimony, is precipitated by addition of a large bulk of water; the mixture is boiled until the precipitate is nearly re-dissolved, more water is added, and the mixture boiled again in like manner, and then filtered. The clear liquid is heated to the boiling point, and then perfectly clean articles of brass are immersed in it. They at once acquire a film of antimony and a lilac colour, and, by allowing them to remain a greater or less length of time, different tints of colour are obtained.

I have observed that zinc readily deposits antimony as a black powder by simple immersion in an aqueous solution of the mixed fluorides of antimony and potassium; that copper also deposits it as a black film and powder by contact with the acid hydrochlorate of terchloride of antimony, and that crystals of silicon did not become coated with antimony in an aqueous solution of terfluoride of antimony containing free hydrofluoric acid; also, that the oxide of iron upon a rusty iron wire was rapidly dissolved in a mixture of equal measures of solution of terchloride of antimony, and a saturated solution of sal-ammoniac. Watt coats copper with antimony by immersing it during about half an hour in a solution of one ounce of chloride of antimony, one pint of spirit of wine, with sufficient hydrochloric acid added to make the mixture clear. I have noticed that antimony is deposited by simple immersion from its ordinary chloride, as prepared for pharmaceutical purposes by zinc, bismuth, tin, lead, brass, and German silver, but not by antimony, nickel, silver, gold, or platinum. According to Raoult, magnesium sets free antimoniuretted hydrogen, but no metallic antimony from solutions of the metal (*Chem. News*, Vol. XIV., p. 27.) Gold, in contact with antimony, in a cold or hot solution of a salt of that metal, does not acquire a metallic coating (*ibid.*, Vol. XL, p. 465).

The electrolysis of antimonic acid and solutions of its salts with electrodes of wood charcoal, or retort carbon, yields stibio-mellogen, and with graphite electrodes stibio-graphitie acid (Bartoli and Papasogli, *Jour. Chem. Soc.*, Vol. XLIV., 1883, p. 592 ; *The Electrician*, Vol. XI., pp. 28 and 101).

A very good solution for obtaining the pure metal by the separate current process, with an anode of antimony, is composed of—

	Parts by weight.
Distilled water	12
Pure hydrochloric acid	$1\frac{1}{2}$
Tartaric acid	1
Potassio tartrate of antimony	1

The electric current should be from about two Smee elements, quite feeble, and of such a strength as to deposit a thickness of metal not exceeding $\frac{1}{32}$nd of an inch per week. The metal thus deposited is hard, close-grained, of a slate-grey colour, silky lustre, and of decided crystalline structure. During deposition, when it has attained a thickness of $\frac{1}{10}$th of an inch, it sometimes cracks spontaneously, and becomes curved in fantastic shapes, or if deposited on a thin metal cathode it causes the latter to bend.

Electrolysis of Teroxide of Antimony.—SbO_3. Molecular weight = 168. This compound, in a fused state, is reduced to metal by contact with charcoal.

Electrolysis of Terfluoride of Antimony.—SbF_3. Molecular weight = 177. This is a very soluble salt, and, unlike the chloride, is not at all decomposed by the addition to it of a large quantity of water. By electrolysis with a separate current of suitable strength, and an anode of antimony, it slowly yields a thick layer of the pure hard grey metal.

By employing a dilute solution of the fluoride containing free hydrofluoric acid, and using a current from two Smee cells, or by passing a current from ten such cells through a saturated neutral solution of the fluoride, during a long period of time, I have obtained very beautiful collections of shining grey crystals of the metal, which do not oxidise by exposure to air.

Electrolysis of Terchloride of Antimony.—$SbCl_3$. Molecular weight = 226·5. From a solution of this salt containing free hydrochloric acid, the antimony may be obtained by the separate current process, and an antimony anode, either in the form of the pure grey metal, or in that known as "amorphous" or "explosive" antimony, according to the degree of density of the current and the composition and temperature of the liquid.

The acidified aqueous solution of chloride of antimony is an excellent conductor of the current; it dissolves an antimony anode freely, yields plenty of the amorphous metal, and does not deteriorate by use or by exposure to the atmosphere. It is, however, decomposed with greater or less rapidity by contact with zinc, cadmium, tin, lead, iron, brass, copper and German silver, each of which deposits the metal upon itself and dissolves. It is also decomposed by water, and therefore articles wet with water must not be immersed in it, and those taken from the liquid must be washed with dilute hydrochloric acid or a solution of tartaric acid previous to washing them with water.

"Explosive Antimony."—To obtain the explosive variety of metal from the usual acidified solution, the current should be of such a strength as to deposit not less than half a grain of metal per square inch of cathode per hour. If the strength is much less, the kind of deposit suddenly changes to the grey variety, sometimes preceded by formation of nodules of that kind upon parts of the surface of the cathode. The two kinds of deposit do not adhere firmly to each other.

A good solution for yielding the "explosive" variety is composed of one ounce of freshly precipitated teroxide or oxychloride of antimony, dissolved in five or six ounces of pure hydrochloric acid of specific gravity 1·12, or it may be made by saturating two volumes of the acid with the oxide or oxychloride, and then adding one additional volume of the acid. The oxide employed should not be that which has been made by oxidising antimony with nitric acid, or with any mixture containing that acid, nor should it be that which has long been exposed to the air. A good solution may also be made by mixing together two ounces of water, four of hydrochloric acid, and eight of finely powdered potassio tartrate of antimony (*i.e.*, tartar emetic). Either of these mixtures will bear a very strong electric current without causing the deposition of a black powder. It yields its metal rapidly, and coatings of any desired thickness may be obtained. I have had some of quite half an inch in thickness. Deposits of one-tenth of an inch thick may be obtained from it in about seventy hours by means of a current from two or three Smee elements.

A suitable depositing liquid may also be prepared by the "battery process," *i.e.*, by immersing a large anode of antimony and a smaller cathode of platinum, silver, or copper in the dilute acid, and passing a copious current until the metal is freely deposited.

The explosive variety of deposit has quite a different appearance from that of the other. It is highly smooth and lustrous, and of a steel-black colour. Its appearance, however, varies somewhat with the speed of deposition. It has the remarkable property that if struck, broken, or rubbed, whilst at the ordinary temperature, or if touched with a red-hot wire, it suddenly rises in temperature, usually about 650°F.; the amount of heat, however, varies with different circumstances. Another difference between the pure variety of metal and the explosive kind is that, when deposited upon a cathode of mercury, the former is absorbed by and alloys with the mercury, but the latter does not.

Like most electro-deposited metals, and especially the pure variety of antimony, the outer and inner surfaces of deposits of the explosive kind are in states of different cohesive strain, which sometimes cause the deposit to crack spontaneously in the depositing liquid and shatter to bits, evolving all its heat. Faint crackling sounds not unfrequently issue from both varieties of the depositing metals whilst forming in the liquids.

Sometimes, with a very dense solution of the chloride, worked rapidly, and the liquid and cathode not at all disturbed, a layer of deposited metal, in the shape of a large button $1\frac{1}{2}$ inch diameter, gradually formed round the cathode, just at the surface of the liquid.

A cylindrical bar of the explosive variety, about $\frac{5}{8}$ths of an inch in diameter, formed upon a rod of tin $\frac{1}{8}$th of an inch thick, when discharged by momentary contact of a heated wire, instantly evolved sufficient heat to melt the tin completely, and the tin ran out and remained liquid a short time. The change which takes place is propagated from particle to particle of the mass. By forming deposits of sufficient thickness upon helices of stout copper wire, and discharging them by application of heat to one end, the change was gradually transmitted to the opposite end, and the velocity of its propagation varied from 12 to 30 feet per minute, the velocity being greater the thicker and more freshly formed and perfect the deposit, and the weaker the cooling influences.

The temperature to which the deposit must be raised in order to produce a sudden discharge varies according to several circumstances, but is usually about 200° or 210°F. in an air bath; the heat, however, of the substance begins to discharge when the metal is heated to between 170° and 190°F., and if a piece of the substance is kept in water maintained at a temperature of 190° to 200°F. during one or two hours, it gradually discharges the whole of its heat. By careful manipulation, keeping one end of a rod of the substance hot and the other cold, the former end only loses its singular property. By carefully breaking thin pieces to small bits between surfaces of wood in ice-cold water, and then triturating them very carefully under such water in a mortar, the active metal may be obtained in a state of powder, with its heat-giving power undiminished. The thermic property gradually diminished by lapse of time, the powdered substance losing it the most quickly; that of a massive deposit disappears in a period of one or two years.

The specific gravity of the pure crystalline variety of deposit varied from 6·369 to 6·673, whilst that of the explosive kind varied from 5·739 to 5·944. Their electro-chemical equivalents also were different, and were determined by electrolysing their solutions in a single undivided circuit with one of cupric sulphate, and weighing the three deposits. I then found from 42·30 to 43·81 parts of the active kind, and 40·41 to 40·79 parts of the grey variety for each 31·7 parts of copper.

The discharge of heat in the explosive kind, whether sudden or gradual, is attended by alterations of the substance. From a bright black, highly lustrous surface, like that of iodine or crystals of silicon, and a bright vitreous-looking fractured surface, it changes to a dull grey earthy appearance and granular fracture. Its cohesive power also changes; and it acquires a strongly acid taste.

As the sudden discharge was attended by evolution of an acid fume (an effect of the heat), and the substance usually lost about 3·5 per cent. in weight, consisting chiefly of chloride of antimony, I made two analyses of the freshly-formed substance deposited from a pure solution of the chloride. The following were the results:—

	No. 1.		No. 2.	
Sb	93·36		93 51	
$SbCl_3$	5·98	} 6·44	6·03	} 6·24
HCl	0·46		0·21	
	99·80		99·75	

·ʳ It appears, therefore, that the deposit is a species of chemical compound of the metal with the ingredients of the liquid ; and that during the change its state of chemical union is destroyed. R. Böttger has stated (*Chemisches Central Blatt*, 1875, p. 674) that the freshly deposited active metal contains occluded hydrogen ; but this has been contradicted.

Smee appears to have been the first to deposit this variety of antimony, but not to notice its singular property ; but since I first observed it in October, 1854, several persons have rediscovered it (see *Comptes Rendus*, Vol. LXXXIII., pp. 854—857; also *Dingl. Poly. Jour.*, Vol. 207, p. 427, and *Jour. Chem. Soc.*, Vol. XL, p. 1,007).

A solution composed of the double chloride of antimony and ammonium, with free hydrochloric acid, may be used instead of that of the acidulated simple chloride for depositing the metal, but possesses no very great advantages.

Electrolysis of Terbromide of Antimony.—SbBr$_3$. Molecular weight = 360. The electrolytic properties of a solution of this salt are much like those of the chloride. It, however, less readily yields a firm deposit of metal. A second variety of the active substance was obtained from it in the following manner:—Dissolve one part by weight of freshly-made teroxide of antimony in ten parts of aqueous hydrobromic acid of specific gravity about 1·3. Filter the solution and electrolyse it with an anode of antimony and a current from three Smee cells, at a speed of deposition of about 4 grains of metal per square inch of cathode per hour.

The deposits thus obtained were of a lighter colour than those from the chloride, they were also quite dull in aspect, and frequently perforated with holes all over the surface like a sponge. This was caused by numerous bubbles of gas. The deposit is less apt to spontaneously crack than the first variety; it is also much more fragile and less hard. Its specific gravity at 60°F. is also much less, and varies from 5·415 to 5·472. It contains a less percentage, viz., 79·52, of metallic antimony. The residue consists of a colourless soft substance composed of terbromide of the metal, and a little hydrobromic acid and water.

It exhibited the same kind of thermic action as the first kind, but the change did not spread throughout the mass unless the substance was previously heated to about 250°F. By contact with a red-hot wire it then evolved all its heat instantly with explosive violence, and with fracture and dispersion of the substance. By gradually heating the entire substance to about 320°F. it exploded suddenly.

In two pairs of experiments made to ascertain the electrochemical equivalent of this deposit I obtained 50·09 and 50·11, also 51·2 and 51·4 parts for every 42·5 parts of the active chloride variety deposited, or 32·2 parts of zinc consumed in the same circuit. Each of the quantities of the two kinds of deposit contained about the same, viz., 40 parts, or ⅓rd of an atomic weight of metallic antimony, the remainder being the associated salt of the metal. These results indicate that in each case the metal alone is deposited by the current; and that during the act of deposition it occludes the saline matter, and acquires the peculiar property.

Electrolysis of Teriodide of Antimony.—SbI_3. Molecular weight = 503. The solution employed was prepared as follows:—Dissolve one part by weight of recently precipitated teroxide of antimony in fifteen parts of aqueous hydriodic acid of specific gravity 1·25. A current sufficiently strong was passed through it, by means of an anode of antimony, to deposit the metal at a rate not exceeding one grain weight per square inch of cathode per hour. During the process the tendency to evolution of hydrogen was so great as frequently to completely disintegrate the deposit.

The substance thus obtained was dull in appearance, grey in colour, scaly, extremely fragile, soft, and much less metallic than even that obtained in the bromide solution. Its specific gravity was 5·27. On immersing it in water bubbles of gas issued from all parts of its surface during a few seconds, and produced a hissing sound like that of slaking lime. Pieces one-ninth of an inch in thickness required to be heated to 338°F. before the contact of a red hot wire would cause them to discharge their heat; they then discharged feebly and evolved red vapours of antimonic iodide.

The unchanged substance yielded on analysis 77·76 per cent. of metal, the remainder being solid red iodide of antimony and a little aqueous hydriodic acid. By depositing it slowly, i.e., at the rate of ·5 grain per square inch of cathode per hour, in the same circuit as the chloride variety, its electro-chemical equivalent was determined, and 48·07 parts were obtained for every 42·5 parts of the other kind. The deficiency of equivalent of metal consisted, no doubt, of deposited hydrogen.

For a more complete account of the several varieties of electro-deposited antimony from the three salts see Phil. Trans. Roy. Soc., 1857, 1858, and 1862; also Chem. News, Vol. VIII., pp. 257 and 281; and Jour. Chem. Soc.

Electrolysis of Tersulphide of Antimony.—SbS_3. Molecular weight = 218. This compound when in a fused state is readily decomposed, and the antimony separated by charcoal, and various metals, e.g., potassium, sodium, copper, tin, iron, &c.

Electrolysis of Chlorides of Arsenic and Antimony.—In the electrolysis of the chlorides of arsenic and antimony some arsenide and antimonide of hydrogen are produced at the cathode. If the three metals—arsenic, antimony, and tin—are simultaneously present they are deposited in the order given. From the solutions of the sulphides of those metals in alkaline sulphides tin and antimony are deposited completely, and arsenic not quite completely, in the metallic state (C. Luckow, *Jour. Chem. Soc.*, Vol. XXXVIII., 1880, p. 283).

Electrolysis of Antimoniate of Potassium.—Bartoli and Papasogli electrolysed an aqueous solution of this salt by means of a current from eight Bunsen cells and an anode of wood charcoal or gas carbon. Very much gas was evolved at the cathode, and a small amount at the anode. The anode was strongly attacked, and the electrolyte became deep black. The authors found in the electrolyte a compound of carbon, hydrogen, oxygen, and antimony, which they term stibiomellogen (*Jour. Chem. Soc.*, Vol. XLII., 1882, p. 406).

For the electro-chemical analysis of antimony compounds see *Jour. Chem. Soc.*, Vol. XLII., 1882, p. 1,320 ; *Chem. News*, Vol. XLVI., p. 106.

Separation of Bismuth.—Bi. Atomic weight = 210. A triad cation. Much less easily deposited in a coherent state than antimony. By simple immersion tin coats itself with very small shining plates of bismuth in a solution of ten grains of nitrate of bismuth and a wineglassful of distilled water, to which two drops of nitric acid have been added. Commaille (*Chem. News*, Vol. XIV., p. 188) states that magnesium deposits bismuth from solutions of its salts by simple immersion. I have observed that it is also deposited from its aqueous chloride by zinc, tin, lead and iron, but not by bismuth, antimony, copper, brass, German silver, gold, or platinum.

Electrolysis of Oxide of Bismuth.—Burckhard states that fused oxide of bismuth is easily decomposed by a current from twelve Bunsen cells with electrodes of copper; if platinum wires are used, an easily fusible alloy of the two metals is formed (*Zeitschrift für Chemie*, Vol. VI., p. 212). Melted oxide of bismuth is instantly reduced to metal by contact with antimony (H. Tamm, *Chem. News*, Vol. XXV., pp. 85–100).

Formation of Peroxide of Bismuth.—Wernicke has prepared this compound as a black deposit, having the composition represented by BiO_2, H_2O, and a specific gravity of 5·571, by electrolysing by a current from two Daniell cells, with an anode and cathode of sheet platinum, a solution of a mixture of basic nitrate of bismuth and tartrate of sodium (*Pogg. Annalen*, Vol. CXLI., p. 109).

Electrolysis of Nitrate of Bismuth.—Sodium amalgam decomposes a saturated solution of nitrate of bismuth, setting free hydrogen and black powder of bismuth (Böttger). By the separate current process an anode of the metal, an extremely feeble current, and a solution of the nitrate in water, with the minimum amount of free acid to render it clear, I have deposited the metal as a very beautiful but thin coating, white, with a faintly pinkish tint, and a fine silky lustre. According to some writers such a deposit is explosive.

Electrolysis of Fluoride of Bismuth.—With pure dilute hydrofluoric acid, a bismuth anode, and a current from a single Smee element, the conduction was extremely feeble, and only a black film was deposited upon a copper cathode in thirty hours.

Electrolysis of Chloride and Iodide of Bismuth.—Metallic bismuth may be deposited upon copper or brass, by means of a current from a single Bunsen cell, from a solution composed of 25 to 30 grammes of the double chloride of bismuth and ammonium, dissolved in a litre of water, faintly acidified with hydrochloric acid. The deposit consists of a blackish mud, with a film of bright adherent bismuth beneath (Bertrand, *Athenæum*, April 22, 1876, p. 570 ; also *Jour. Chem. Soc.*, Vol. I., 1876, p. 451). I have deposited the metal, evidently containing some ingredients of the liquid, by a separate current and a bismuth anode, from a solution of iodide of bismuth and iodide of potassium. The deposit was an extremely bulky, jet black powder, which contained iodine even after most persistent washing, and became slowly oxidised by exposure to the atmosphere.

A cyanide solution has also been recommended for depositing bismuth, but an anode of that metal does not readily dissolve in a hot solution of potassic cyanide.

One part of bismuth in 1,200,000 parts of mercury may be detected by the addition of potassium amalgam and water, the bismuth being electrolytically separated as a black powder on the sides of the vessel (Serullas, *Ann. Chem. et Phys.*, 3rd Series, Vol. XXXIV., p. 192).

For the electrolytic analysis of compounds of bismuth see V. Francken, *Chem. News*, Vol. XLVI., p. 106 ; also *Jour. Chem. Soc.*, Vol. XLII., 1882, pp. 896 and 1,320.

Separation of Osmium.—Os. Atomic weight = 199. A cation. Zinc deposits osmium in the form of black flocks when immersed in a solution of the black oxide in concentrated hydrochloric acid. Metallic mercury decomposes a solution of osmic acid, and forms an amalgam of osmium and mercury (Tennant).

Smee electrolysed a solution of osmic acid (OsO_4), and obtained a black deposit. Wöhler passed a current from two Bunsen cells by means of an anode of osmium through dilute sulphuric acid; the metal was freely converted into osmic acid ; also through a solution of caustic soda the latter became of a deep yellow colour, and a deposit of osmium was formed upon the cathode (*Chem. News*, Vol. XIX., p. 10).

Ruthenium.—Ru. Atomic weight = 104·2. A cation. No reliable electrolytic experiments appear to have been made with this metal. Its great rarity and infusibility, extreme cost, and porous structure, are the chief obstacles.

Separation of Rhodium.—Ro. Atomic weight = 104·3. A cation. Smee stated that by means of a separate current from ten of his cells, and platinum electrodes, he deposited this metal from a solution of its sodio-chloride, and obtained a brittle white deposit; and that with a stronger current the deposit was a black powder.

Separation of Iridium.—Ir. Atomic weight = 197. A cation. Sodium amalgam decomposes a concentrated solution of sodio-iridium chloride, and forms an amalgam of iridium. According to F. Wöhler, osmi-iridium is readily dissolved as an anode in a solution of caustic soda. Smee stated that he had deposited this metal in a bright reguline state on a small scale. According to a writer in *Dingler's Polytechnick Journal*, both electro-deposited iridium and rhodium detonate when heated (*Jour. Chem. Soc.*, 2nd Series, Vol. XL, p. 1,007), probably in consequence of their containing hydrogen.

Separation of Palladium.—Pd. Atomic weight = 106·5. A cation. Mercury decomposes a solution of a palladium salt, and forms an amalgam. According to S. Kern (*Chem. News*, Vol. XXXIII., p. 236), the immersion of magnesium in aqueous solutions of salts of palladium yields hydrogen, monoxide of palladium, and hydrogenated palladium.

Formation of Peroxide of Palladium.—PdO_2. Molecular weight = 158·5. An anode of palladium, used to conduct a current from two Bunsen elements into dilute sulphuric acid, became slowly covered with an almost insoluble film of this compound (F. Wöhler, *Chem. News*, Vol. XIX., p. 10).

I passed a current from fifty Smee elements, by means of a palladium anode and platinum cathode, through dilute sulphuric acid. No odour of ozone occurred at the anode, unless the latter dipped but a very small depth into the liquid. Conduction was copious, and a deposit of splendid colour—red, purple, &c.—formed upon the anode. By reversing the direction of the current the now well-known phenomenon of bending of the cathode by absorption of hydrogen took place; and by removing the charged cathode from the liquid and bending it by mechanical means it evolved heat.

Electrolysis of Nitrate of Palladium.—A solution of this salt is said to be a good conductor, but apt to yield the metal in the form of a black powder. I electrolysed strong nitric acid, by means of a current from fifty Smee cells, with a palladium anode and platinum cathode. Copious conduction, and rapid decomposition of the acid, with abundant evolution of red fumes, took place. Much gas was set free from the anode, but none from the cathode until after a short time. The anode was not at first visibly corroded, but after half an hour's action the palladium slowly dissolved, forming a red liquid. No metallic deposit formed upon the cathode.

Nitrate of palladium dissolved in water and acidified with a little nitric acid, deposited upon the cathode a bronze coloured coating, which by prolonged action became darker, and then black, and was easily soluble in nitric acid. Some reddish oxide was formed upon the anode. Alkaline solutions behaved similarly, except that the action was slower, and the deposited metal more adhesive (Schucht. Berg ünd Hüttenmannische Zeitung, *Chem. News*, Vol. XLI., p. 280).

Formation of Fluoride of Palladium.—PdF_4.

Molecular weight $= 182 \cdot 5$. I electrolysed thirty per cent. pure aqueous hydrofluoric acid, with a sheet palladium anode, in a large platinum vessel as the cathode, by means of a current from six Smee elements. Free conduction occurred, and much gas was evolved from each electrode, and there was a strong odour of ozone. A dark, red-brown, thin coating for med upon the palladium, but did not dissolve during fifteen hours' electrolysis. The liquid was filled with minute floating particles of palladium, caused by the reducing action of the hydrogen from the cathode. After six days' continuous action the anode was greatly corroded.

I also electrolysed the pure anhydrous acid in a chilled state, in the same vessel, with a thick palladium anode and the same current, and sometimes with a current from thirty cells. The process was difficult and very dangerous, and notwithstanding the low temperature, and that the vessel was closely covered by a lid of paraffin, the acid volatilised rapidly and fumed greatly, partly in consequence of the escaping hydrogen and the heat of conduction resistance. The coldness of the vessel and the intense attraction of the acid for moisture caused water to condense upon the lid of the vessel, and rendered it difficult to preserve the acid in the anhydrous state. The lid was therefore made to overhang the edge of the cup, and was also covered by a layer of cotton wool. With a current from twenty cells the conduction was copious. The anode quickly became coated with a thick, dark, red-brown, brittle crust, which was of a redder colour on the side next the anode, and did not perceptibly impede the passage of the current. This crust was scraped off at intervals of about one hour into a platinum dish standing upon a slab of iron heated to about 350°F., and at once transferred to a closed platinum bottle.

After eleven hours' action, the acid was still colourless, as if the crust upon the anode was perfectly insoluble. Some black powder, which proved to be metallic palladium, was, however, found upon the cathode, and indicated that some palladium had dissolved and been reduced. The crust also on the side towards the anode was nearly black when dry, and showed signs of metallic particles when pressed between smooth surfaces of agate, indicating some reduction by the diffused hydrogen.

In some other experiments the hydrogen was more perfectly excluded from touching the anode. The platinum cup was 2¾ inches wide and 3¼ inches deep, divided into two equal parts by a well-fitting vertical plate of paraffin, which extended to within half an inch of the bottom. The palladium anode and platinum cathode were each about 4 inches long and 1 inch wide, and firmly fixed in slits in the two halves of the paraffin lid. With 5½ ounces of the perfectly anhydrous acid, and a current from twelve one-pint Grove cells, the conduction was copious, and in five minutes the immersed part of the anode had acquired a deep brown colour. The electrolysis was continued during five hours, the anode being taken out and scraped each half hour, and the crust preserved in a platinum bottle. The crust was hard, and sparks were sometimes caused by particles of it being decomposed by the heat of friction in removing it. A hissing sound was heard during the whole of the electrolysis, but the density of the fumes prevented any effervescence being seen. 10·46 grains of black powder were found upon the cathode and adjacent parts of the vessel and partition, and yielded 10·11 grains of metallic palladium; this indicated some small degree of solubility of the crust, and the great necessity of excluding hydrogen from touching the anode. The anode had lost 37·90 grains in weight, and 54·13 grains of the dry brown crust was obtained. After deducting the 10·11 grains of palladium found in the liquid, the remaining quantity of corroded metal would form only 47·62 grains of tetrafluoride ; the crust therefore contained in addition probably some hydrofluoric acid.

I also found that a palladium anode was very rapidly corroded by the passage of a current from six Grove elements through pure potassic fluoride in a state of fusion. Finely divided palladium was found in the saline residue.

Electrolysis of Chloride and Iodide of Palladium.—I electrolysed concentrated, also dilute hydrochloric acid, by a current from fifty Smee cells with a palladium anode and a platinum cathode. Action was instant and rapid, hydrogen was copiously evolved from the cathode and chlorine from the anode, and the anode dissolved, forming a blood-red liquid, and a black deposit of palladium soon formed upon the cathode.

M. A. Bertrand recommends a perfectly neutral solution of the double chlorides of palladium and ammonium, with or without the use of a separate current, for the separation of this metal (*Chem. News,* Vol. XXIV., p. 227). It is stated that with a separate current from two or three cells and a palladium anode, the current is somewhat impeded in this solution by a bright golden yellow coating forming upon the anode. Iodide of palladium dissolved in a solution of potassic iodide is an unsatisfactory one for yielding reguline metal.

Electrolysis of Cyanide of Palladium.—$Pd.Cy_2$. Molecular weight $= 158.5$. A solution of the double cyanide of palladium and potassium, containing free potassic cyanide, is said to yield by the separate current process thick deposits of the metal in a reguline state. The liquid holds in solution a large quantity of the metal, and may be prepared either by the usual chemical means—*i.e.*, by dissolving cyanide of palladium in a solution of potassic cyanide, or by the battery process. A less satisfactory mixture may also be made by dissolving chloride of palladium in a solution of the alkaline cyanide. A solution for depositing palladium is prepared as follows:— Cyanide of palladium is mixed with 7 per cent. of ferrocyanide of potassium, 3 per cent. of caustic potash, and 60 per cent. of water, and the mixture boiled during half an hour, the lost water being replaced (Frantz, *Chem. Centr.*, 1876, p. 592).

Separation of Platinum.—Pt. Atomic weight $= 197$. A tetrad cation. Nearly all electrolytic operations with platinum are performed with a solution of the tetrachloride, or with hydrochloric acid, because, with the exception of anhydrous hydrofluoric acid, a solution which yields chlorine at the anode is nearly the only one in which metallic platinum will dissolve; even a solution of potassic cyanide with a strong current corrodes an anode of platinum but sparingly. Nearly all the ordinary metals become coated with platinum in its solutions by the process of simple immersion, and it may be separated by each of the methods of electrolysis.

Magnesium in a solution of platinic chloride evolves hydrogen, and after about twenty hours deposits a black powder of metallic platinum. On leaving it in contact with water a brown hydrate of platinum is formed (S. Kern, *Jour. Chem. Soc.*, 1876, Part I., p. 684). Sodium amalgam decomposes a concentrated solution of platinic chloride, also one of chloroplatinate of ammonium, and forms in each case an amalgam. According to Joule, the electro-deposition of platinum with a cathode of mercury produces an amalgam of the two metals. I observed that by simple immersion of arsenic, antimony, tellurium, bismuth, zinc, cadmium, tin, lead, iron, cobalt, nickel, copper, brass, German silver, mercury, or silver in a solution of tetrachloride of platinum, they became coated with that metal. According to Lan, a solution of one part of platinic chloride in 15 parts of alcohol and 50 of ether deposits platinum on tin, brass, and white metal (*Jour. Chem. Soc.*, Vol. XLII., 1882, p. 1,145). Böttger adds carbonate of sodium to platinic chloride as long as carbonic anhydride is evolved, then a little starch sugar, and, finally, chloride of sodium till the precipitated platinum appears white. The resulting solution coats articles by simple immersion (Watts's "Dictionary of Chemistry," Vol. VI., p. 950).

One of the best liquids for obtaining thick reguline deposits by the separate current process is that of Roseleur, who prepared it as follows :—Convert 10 parts of platinum into dry tetrachloride, and dissolve it in 500 parts of distilled water (the whole should dissolve). Add, with stirring, to the solution 100 parts of crystallised phosphate of ammonia previously dissolved in 500 parts of distilled water ; and as this produces a precipitate, add at once, with copious stirring, a ready-made solution of 500 parts of crystalline phosphate of soda in 1,000 parts of distilled water. Boil the mixture until an odour of ammonia ceases and the liquid begins to turn blue litmus paper red. The liquid must be used hot, and a strong current must be employed, because an anode of platinum does not dissolve in it. It is decomposed by, and deposits platinum upon, zinc, tin, or lead by simple contact. A solution made by dissolving tetrachloride of platinum in one of potassic cyanide has also been used for the same purpose, but it also does not dissolve a platinum anode.

Formation of Tetrafluoride of Platinum.—In some experiments of mine a platinum anode in water, containing 10 per cent. of pure anhydrous hydrofluoric acid, was not corroded by the passage of a current from six Smee or six Grove elements during many hours. Very free conduction occurred, a powerful odour of ozone and a gas which inflamed a red-hot splint were evolved at the anode, but no platinum was deposited. With an aqueous solution of pure potassic fluoride precisely similar effects occurred.

A current from 24 elements of magnesium and platinum in an exciting solution of common salt was passed during 18 hours, by means of platinum electrodes, through water containing 40 per cent. of the same pure acid, but no corrosion of the anode took place. With a current from ten Smee cells, and platinum electrodes I also electrolysed water containing 80 per cent. of the pure acid. Abundant conduction with evolution of hydrogen and ozone occurred; the anode lost 16·58 grains by corrosion during 36 hours, and became covered with a brownish black crust, which partly dissolved in the liquid to a brownish solution. No platinum was deposited, probably because the hydrogen decomposed the solution. I also electrolysed with platinum electrodes the pure anhydrous acid in a precisely similar way to that described under "Fluoride of Palladium." With a current from forty Smee cells the anode corroded rapidly, and acquired a dark red brown crust, which was insoluble in the acid, but rapidly deliquesced in the air; it dissolved, with partial decomposition, to a basic salt and formation of a blood-red liquid, in water.

By electrolysis with platinum electrodes, during 16 hours, of water containing 40 per cent. of the pure acid, mixed with its own bulk of strong nitric acid, gases were freely evolved; but scarcely any platinum was dissolved, and none was deposited. With an equal bulk of strong hydrochloric acid substituted for the nitric, hydrogen and chlorine were set free, but in four hours' action the anode was very little corroded. With the same volume of sulphuric instead of the nitric acid, after many hours' action, the anode was again but little corroded. With phosphoric anhydride dissolved in the dilute hydrofluoric acid, and the mixture electrolysed, the results were similar. And with much selenious acid dissolved in it, selenium containing traces of platinum was freely deposited, and gas was evolved as before (see *Phil. Trans. Roy. Soc.*, 1869, p. 200).

By electrolysing fused fluoride of potassium or lithium with platinum electrodes, the anode was rapidly dissolved, and the resulting salt of platinum instantly decomposed, and its metal set free; and by electrolysing pure double fluoride of hydrogen and potassium in a fused condition, the platinum anode was rapidly dissolved, and a colour imparted to the salt. Fused silico-fluoride of potassium, or the fused fluorides of silver, copper, lead, manganese, or uranium, when electrolysed by a current from six Smee cells, did not corrode a platinum anode.

Electrolysis of Tetrachloride of Platinum.— Pt. Cl_4. Molecular weight = 339. This salt may be formed by the electrolysis of hydrochloric acid with a platinum anode and a dense current. According to Commaille (*Chem. News*, Vol. XIV., p. 188) magnesium deposits pure platinum from a solution of platinic chloride. I have observed that crystals of silicon did not acquire a coating of platinum in that liquid. A smooth deposit of platinum upon bright copper may be obtained by immersing the copper in a boiling solution composed of 100 parts of distilled water, 12 of caustic soda (or 40 of sodic carbonate), and 10 of platinic chloride. Copper and brass may also be coated by means of contact with zinc in a solution prepared as follows :—To a solution of platinic chloride add sodic carbonate in fine powder until effervescence ceases, then add some glucose, and afterwards as much sodic chloride as will produce a white precipitate. The solution should be used at a temperature of 60°C. ("Les Mondes," *Chem. News*, Vol. XIX., p. 226).

Separation of Gold.—Au. Atomic weight = 196·6. A triad cation. Like platinum, gold is very easily separated from its solutions by each of the methods of electrolysis. Thallium deposits gold from its solutions (W. C. Reid, *Chem. News*, Vol. XII., p. 242). Auric terchloride and the auro-cyanide of potassium are the only common soluble salts of the metal; a solution of the oxide in hydrobromic acid, or of the aurate of ammonia (a very explosive substance) in potassic cyanide previously dissolved in water, may also be employed. A gold anode is corroded and dissolved in various liquids, *e.g.*, hydrochloric acid, solution of sodic chloride, &c. Runspaden has observed that a gold anode in dilute sulphuric acid is considerably oxidised, and a definite hydrated oxide of gold is formed (*Chem. News*, Vol. XX., p. 179). By immersing zinc in a solution of sulphide of gold dissolved in one of sulphide of ammonium, excluded from the atmosphere, it becomes coated with gold (C. D. Braun, *Chem. News*, Vol. XXIX., p. 230).

J. Schiel has produced Nobili's rings on a horizontal plate of burnished gold used as the anode in very dilute nitric acid, the negative pole being a platinum wire, supported a short distance above the gold plate. After passing a current of suitable strength during about ten minutes, the plate was washed, dried, and exposed a few hours to sunlight; the rings then appeared of brilliant colours. With an alkaline solution the effects were less powerful (*Pogg. Ann.* CLIX., p. 493).

Under the influence of an electric current nitric acid dissolves gold (Berthelot, *Jour. Chem. Soc.*, Vol. XXXVIII., 1880, p. 158).

Formation of Fluoride of Gold.—By electrolysing pure dilute hydrofluoric acid with a gold anode, in a platinum crucible as the cathode, by means of a current from six Smee cells, during many hours, the current passed very freely, much gas came from each electrode, and an odour of ozone from the anode. The anode gradually became covered with an insoluble red-brown film. None of the metal dissolved. Similar effects occurred with more concentrated acid, and with a very much stronger current. The red-brown film appeared to be gold; it was insoluble in nitric acid, and when burnished with agate it appeared like gold.

With a gold anode in pure anhydrous hydrofluoric acid, at 10°F., even a current from forty Smee cells was but feebly transmitted. In one and a-half hour the anode acquired a dark, reddish-brown film, with a few crystals, at first of a

greenish colour, upon its edges. By exposure to the air the
crystals became first yellow and then red. A current from
six Smee cells was conducted freely by a gold anode, in a
solution of pure fluoride of ammonium containing free ammonia.
The anode evolved much gas, and became covered with an
insoluble, bright, lemon-coloured powder, but no gold appeared
on the cathode. By means of a current from three and also
from six Grove elements I electrolysed with a gold anode the
pure fluorides of potassium and lithium in a melted state.
Metallic gold separated, and the anode was very rapidly
corroded.

Electrolysis of Auric Terchloride.—$Au.Cl_3$. Molecular
weight = 303·1. Sodium amalgam easily reduces a solution
of auric terchloride, and, according to G. A. Koenig, even charcoal reduces it to metal by simple immersion (*Journal* of the
Franklin Institute, May, 1882; see also *Chem. News*, Vol.
XLIV., p. 215). The mere contact of magnesium, phosphorus,
arsenic, antimony, tellurium, bismuth, palladium, silver,
mercury, copper, and nearly all the base and brittle metals,
with this solution separates the metal. I have noticed that
crystals of silicon did not reduce it, but that by contact of
amylene, "petroleum ether," benzine, coal gas, and numerous
liquid hydrocarbons, with the aqueous solution, films of the
metal gradually separated (see *Proc. Birm. Phil. Soc.*, Vol. IV.,
Part I.).

According to D. Tommasi, solution of auric chloride is not
reduced to metal by hydrogen or platinum alone, but only by
hydrogen in the presence of platinum (*Chem. News*, Vol. XLI.,
p. 116).

Electrolysis of Auro-Cyanide of Potassium.—$KCy.AuCy$.
Molecular weight = 287·7. In a solution of the double cyanide
of gold and potassium, zinc, copper, brass, and German silver,
became gilded by simple immersion ; but platinum, gold, silver,
nickel, iron, lead, tin, bismuth, and antimony did not.

This salt, when dissolved in a suitable proportion of water,
and a certain proportion of potassic cyanide added, constitutes
the ordinary electro-gilding solution. The compound may
either be formed by dissolving the salts in water, or by taking
a solution of potassic cyanide, and electrolysing it with an
anode of gold and a cathode of platinum, until gold is freely
deposited. This process, however, leaves a large excess of
simple potassic cyanide in the liquid, and also by abstracting
some of the cyanogen to form auric cyanide, and substituting
oxygen in its stead, it introduces caustic potash, and the
caustic alkali gradually becomes carbonate by absorbing carbonic acid from the atmosphere. The solution, when formed,
yields by electrolysis gold at the cathode, whilst cyanide of
gold is formed at the anode, and dissolves.

For deposition of gold by an electric current in solution of potassic ferrocyanide and auric chloride, see E. Ebermayer, *Jour. Chem. Soc.*, Vol. XXXIV., 1878, p. 178. For a solution suitable for gilding iron, see Watts's "Dictionary of Chemistry," Vol. VIII., Part II., p. 1,119. A great variety of mixtures, containing auric chloride or cyanide, and other substances, such as the carbonates and chlorides of potassium and sodium, sodic bisulphate, phosphate, and pyrophosphate, potassic ferrocyanide, and sulphocyanide, aqueous ammonia, carbonate of ammonium, &c., have been employed for electro-gilding. The particulars of their composition may be found in "The Art of Electro-Metallurgy," Longmans and Co.'s "Text Books of Science."

By electrolysing a solution of methylamine with a gold anode and a feeble current during several days I obtained no deposit of gold.

Separation of Silver.—Ag. Atomic weight = 108. A monad cation. Its commonest soluble salts are the nitrate, acetate, argento-potassic cyanide, ammonio-nitrate, and ammonio-chloride. Other soluble ones are ammonio carbonate, sodio hyposulphite, argento-potassic iodide, potassio-tartrate, and argento-potassic sulphocyanide. Sulphate of silver is but slightly soluble. All the solutions of silver are readily decomposed by an electric current with deposition of metal upon the cathode, and in some cases with oxidation of the silver of the liquid at the anode, and formation of argentic peroxide. Nearly all the solutions of the salts of silver are reduced to metal by simple immersion in them of any of the base metals. Aluminium reduces the silver from an ammoniacal solution of argentic chromate (Watts's "Dictionary of Chemistry," Vol. VII., p. 54). According to Tribe, silver deposited by copper always contains copper, if the solution has absorbed air (*Chem. News*, Vol. XXIV., p. 76). Gold in contact with silver in a cold or hot acid or neutral solution of a salt of silver receives no deposit of silver (Raoult, *Jour. Chem. Soc.*, Vol. XI., p. 465).

Formation of Silver Peroxide.—F. Wöhler states that if a current from two Bunsen cells is passed through a dilute solution of sodic sulphate or dilute sulphuric acid, by means of a silver anode, the latter becomes coated with argentic peroxide, due to the action of ozone. With a solution of potassic nitrate similarly treated, brown argentic oxide is formed; with one of potassic ferrocyanide the anode becomes coated with white amorphous argentic ferrocyanide, and with one of potassic bichromate it becomes covered with reddish-black crystallised argentic chromate (*Chem. News*, Vol. XVIII., p. 189). Brester states that by electrolysing melted caustic soda, with an anode of silver, the anode dissolved and silver was deposited on the platinum cathode, and that on cleansing the cathode with nitric acid, a residue of black powder of platinum was obtained (*Chem. News*, Vol. XVIII., p. 145). I have on various occasions noticed a similar residue after electrolysing melted argentic fluoride with platinum electrodes.

Electrolysis of Argentic Nitrate.—$Ag.NO_3$. Molecular weight = 170. According to Brester, hydrogen, evolved either by electrolysis, by the decomposition of steam by red-hot iron, or by zinc or iron in dilute sulphuric acid, reduces a solution of argentic nitrate, but not one of the sulphate; also, if a cathode of platinum, whilst being used in the electrolysis of dilute sulphuric acid, be instantly dipped into a solution of the nitrate, it sometimes reduces the silver and sometimes not (*Chem. News*, Vol. XVIII., p. 144). Russel observed that pure hydrogen reduces a solution of the nitrate to metal and nitrite (Watts's "Dictionary of Chemistry," Vol. VIII., Part II., p. 1,070).

Sodium amalgam decomposes a strong solution of argentic nitrate, and forms an amalgam of silver and mercury. Joule formed the same amalgam, but richer in silver, by depositing silver from the same solution into a cathode of mercury. According to W. C. Reid, thallium deposits silver from a solution of its nitrate by simple immersion (*Chem. News*, Vol. XII., p. 242). Metallic mercury, in an acidified and moderately strong solution of the same salt, forms a "silver tree" or "Arbor Dianæ." I observed that an aqueous solution of argentic nitrate yielded its metal by simple immersion to arsenic, antimony, bismuth, mercury, copper, brass, German silver, nickel, iron, lead, tin, cadmium, and zinc, but not to silver, gold, or platinum. In an alcoholic solution of the salt, antimony, bismuth, zinc, tin, copper, brass, and the alloys of silver with zinc, tin, or lead, deposited silver by simple immersion, but iron did not.

Electro-deposited nickel does not separate silver by simple immersion from a solution of argentic nitrate (J. Spiller, *Chem. News*, Vol. XXIV., p. 175). Aluminium after six hours' immersion begins to precipitate the silver, either from slightly acid or neutral solutions, whether concentrated or dilute (A. Cossa, Watts's "Dictionary of Chemistry," 2nd Supplement, p. 54). According to S. Kern, magnesium precipitates oxide of silver from an aqueous solution of argentic nitrate (*Chem. News*, Vol. XXXII., p. 309).

Fused argentic nitrate, when electrolysed by a separate current, yields silver at the cathode and a large amount of oxygen at the anode.

Formation of Argentic Peroxide.—Ritter, in 1814, discovered that when a concentrated solution of argentic nitrate is electrolysed with two thick platinum wires as electrodes peroxide of silver, Ag_2O_2, is deposited in crystals upon the anode, and metallic silver upon the cathode. Fischer states that these crystals always contain argentic nitrate (Watts's "Dictionary of Chemistry, Vol. V., p. 303). I have also found a nitrogen compound in them.

To deposit coherent silver from a solution of the nitrate requires the liquid to be weak and the current feeble. According to Luckow, the formation of peroxide of silver at the anode in a solution of argentic nitrate may be prevented by the addition of glycerine, milk, sugar, or tartaric acid (*Chem. News*, Vol. XLII., p. 76).

Berthelot obtained sesquioxide of silver by the electrolysis of a 10 per cent. solution of argentic nitrate. It was in the form of large, thick, black, lamillar, striated needles, of brilliant metallic lustre (*Jour. Chem. Soc.*, Vol. XXXVIII., 1880, p. 442).

Electrolysis of Argentic Fluoride.— Ag.F. Molecular weight $= 127$. I have observed that carbon and crystalline boron do not separate silver from fused argentic fluoride at a red heat, but that crystals of silicon thrown upon the melted salt become red hot, and burn vividly, producing silicic fluoride and separating silver ; also that hydrogen separates silver from the semi-fluid salt. In an aqueous solution of the salt, crystals of boron did not separate silver, but crystals of silicon deposited slowly crystals of silver. Stannous fluoride in contact with platinum also separated silver from such a solution. I observed also that in a mixture of solutions of argentic fluoride, hydrofluoric and nitric acids, crystals of silicon evolved spontaneously inflammable bubbles of silicide of hydrogen gas (*Chem. News*, Vol. XX., p. 28, and XXIV., p. 291).

I found the following to be the chemico-electric order of various elementary substances in fused argentic fluoride, the first being the most positive :—Silver, platinum, charcoal of lignum vitæ, palladium, gold. And in a dilute aqueous solution of the salt, aluminium, magnesium, silicon, iridium, rhodium, and carbon of lignum vitæ, platinum, silver, palladium, tellurium, gold (*Chem. News*, Vol. XXI, p. 28).

In a number of experiments of electrolysing argentic fluoride in a fused state in a covered platinum vessel with platinum electrodes, with a current from six Smee cells, conduction commenced before the salt had fused, and when the salt had become quite liquid the conduction appeared to be as perfect as when the electrodes were united by a wire. No signs of genuine electrolysis were observable in either case. I also electrolysed the fused salt by a current from ten Smee cells, with an anode of highly ignited charcoal of lignum vitæ. Very little conduction took place ; the anode was, however, corroded, and evolved gas (*Phil. Trans. Roy. Soc.*, 1870, p. 234; *Chem. News*, Vol. XXL, p. 28).

I also electrolysed pure anhydrous hydrofluoric acid in a chilled state by means of a silver anode and a current from ten Smee cells. Conduction was free, the anode corroded rapidly, and became covered, first with some black powder upon its edges, probably peroxide, and then with a grey powder, probably silver, which contained only a trace of soluble silver salt.

By electrolysing a saturated neutral aqueous solution of argentic fluoride, with a small platinum anode and a large platinum cathode, by a current from six Grove cells, the conduction was free, but no gas or odour was evolved. A thick, hard, and strongly adherent crust of argentic peroxide formed upon the anode. By using a more dilute solution a similar crust was formed, and gas was evolved from the anode. An aqueous solution of this very soluble salt was decomposed by an extremely feeble current with great ease. The deposition of silver, also, with this solution was so rapid that the depositing vessel soon became largely occupied by a loose, bulky mass of fibrous crystals of silver, which soon metallically united the two electrodes if not frequently prevented.

In the electrolysis of solutions of argentic fluoride containing free hydrofluoric acid, with silver anodes, I repeatedly observed that after the current has passed some time the anode becomes extremely brittle and porous, and its surface crumbles away, and falls as a powder to the bottom of the vessel, instead of dissolving smoothly, as with silver, in a solution of potassic cyanide. In order to test whether free fluorine diffused into the metal, I employed as an anode a pure silver tube, closed at the bottom, and communicating at the top with a pressure gauge. No signs of gas or of free fluorine were, however, observed by means of this test, or by chemical ones.

Electrolysis of Argentic Chloride.—Ag.Cl. Molecular weight = 143·5. Aluminium evolves great heat by contact with melted argentic chloride, and separates the silver in melted globules. It also precipitates silver as a fine crystalline powder from an ammoniacal solution of the chloride (A. Cossa, Watts's "Dictionary of Chemistry," Vol. VI., p. 54). Fused argentic chloride is resolved by the current into silver at the cathode and chlorine at the anode (Faraday). Sodium amalgam reduces the chloride, bromide, and iodide of silver, when in contact with water.

Electrolysis of Argentic Chlorate.— Ag.ClO$_3$. Molecular weight = 191·5. A solution of this salt containing free chloric acid was easily decomposed by a current from two Smee cells, with a silver anode. It yielded a copious deposit of loose silver upon the cathode, and upon the anode a black crust, apparently of argentic peroxide, which soon stopped the current. To electrolyse it properly requires a feeble current, a large cathode, and a very large anode (see *Proc. Birm. Phil. Soc.*, Vol. IV., Part I.).

Electrolysis of Silver Perchlorate.—A solution of argentic perchlorate, containing free perchloric acid, with a silver anode, is a remarkably good conductor. It conducted copiously a current from a single Smee element, and was decomposed even more readily than the chlorate. The anode was rapidly corroded, and acquired first a thick loose coating of black solid matter, and then one of a dark green colour. To electrolyse this liquid properly requires a very feeble current, a rather small anode, and a very large cathode (see *Proc. Birm. Phil. Soc.*, Vol. IV., Part I.).

Electrolysis of Argento-sodic Sulphite. — An aqueous solution of this salt is said by Roseleur to possess a singular property. When a piece of metal is immersed in a solution of another one, in which it coats itself with that metal, a portion of the immersed one dissolves, and produces an immense number of minute electric currents which pass from an infinity of minute portions of the surface of the metal into the liquid, decomposing it, and re-enter at other minute portions of the metallic surfaces, and deposit an equivalent weight of the other metal as a coating upon the immersed one ; but in this particular solution a spontaneous *chemical* change also occurs in the liquid itself, the sulphurous anhydride of the argentic sulphite takes oxygen to itself to form sulphuric anhydride, and sets the silver free, and this silver adheres to any solid surfaces present, *i.e.*, to the immersed metal, and to the containing vessel. The sulphuric anhydride unites with some of the soda of the undecomposed portion of the sulphite, and liberates sulphurous anhydride, and forms sulphate and bisulphite of sodium. This action is very similar to that which takes place in certain processes of coating looking-glasses with pure silver.

A solution also composed of sulphite of silver dissolved in an aqueous solution of potassic sulphite has been used for depositing silver by the separate current process. It is a very good one except that it gradually decomposes and deposits its silver by the influence of light. A solution has also been formed by dissolving argentic chloride in an aqueous solution of sodic hyposulphite. It easily yields its metal by electrolysis with a current, but under the influence of light it is decomposed, and its silver precipitated as argentic sulphite.

Electrolysis of Argentic Sulphate.—$Ag.SO_4$. Molecular weight $= 204$. In an aqueous solution of the sulphate of silver, antimony, tin, iron, copper, and the alloys of silver with zinc, tin, or lead, deposited the silver by simple immersion, but bismuth did not.

Electrolysis of Argento-potassic Cyanide.—Ag.Cy.KCy. Molecular weight = 199·1. A solution of this salt, containing an excess of potassic cyanide, constitutes the ordinary silver-plating liquid. It may be formed by dissolving the salt in water, say one or two ounces per gallon—the exact proportion is not material—and then adding about one-tenth of its weight of potassic cyanide. Nearly the same mixture is obtained by electrolysing a solution of potassic cyanide with a silver anode, until silver is freely deposited; in this case, however, caustic potash is formed in the liquid, and gradually becomes converted into carbonate by contact with the air. Electrolysis of the solution yields silver alone at the cathode, and at the anode argentic cyanide, which dissolves. A number of modifications of this liquid have been employed, such as solutions formed by dissolving nitrate, chloride, or ferro-cyanide of silver in one of potassic cyanide, but the above mixture is the best.

By dissolving some bisulphide of carbon in a strong solution of potassic cyanide, and adding *a very minute proportion* of this solution occasionally to the above mixture, the physical character of the deposited silver is greatly changed; instead of being soft and somewhat dull white in appearance, it becomes hard and highly lustrous, like burnished metal. I have found by chemical analysis that it contains a minute proportion of sulphur.

By electrolysing a 33 per cent. aqueous solution of methylamine with a silver anode and a feeble current from a single Smee cell, the anode slowly dissolved, and a loose deposit of silver crystals formed upon the cathode. Somewhat similar results were obtained with a strong solution of trimethylamine.

Blagden states that the desilvering of lead is facilitated by dissolving about half a per cent. of zinc in the refined metal at 540°C., and passing a voltaic current through it by means of copper wires, until all the zinc has risen to the surface! This crust contains the silver, and may be removed after the melted mass has fallen to 450°. The process must be repeated several times.

For a solution fit for silvering iron, *see* Watts's "Dictionary of Chemistry," Vol. VIII., Part II., p. 1,119. For deposition of silver from a pasty mixture of salts by simple contact, *see* Roseleur, *Jour. Chem. Soc.*, Vol. XXXIV., 1878, p. 538. For the electrolytic analysis of silver, see *Chem. News*, Vol. XLII., p. 331 and p. 76 ; Watts's "Dictionary of Chemistry," Vol. VII., p. 790 ; *Jour. Chem. Soc.*, Vol. XXXVIII., 1880, p. 747.

Separation of Mercury.—Hg. Atomic weight = 200. A dyad cation. I have observed that solutions of mercurous salts have their metal deposited by simple immersion, by arsenic, antimony, bismuth, zinc, cadmium, tin, lead, iron, copper, brass, and the alloys of silver with zinc, tin, lead, or copper. Iron deposited mercury from a solution of mercury acetate. A. Cossa states that aluminium deposits mercury by simple immersion in aqueous solutions of mercuric nitrate, chloride, and cyanide; also from a solution of mercuric chloride in alcohol, and of mercuric iodide in one of potassic iodide (Watts's "Dictionary of Chemistry," Vol. VII., p. 54). Thallium deposits mercury from an aqueous solution of mercurous sulphate (W. C. Reid, *Chem. News*, Vol. XII,. p. 242). Solutions of salts of mercury have been electrolysed by the mutual contact of two metals in them (*see* Gladstone and Tribe's experiments, *Phil. Mag.*, 4th series, Vol. XLIX., p. 245).

I have observed that by passing a current from a mercury anode through dilute sulphuric acid into a platinum cathode the latter soon acquires a coating of mercury. E. Obach states that a liquid alloy of sodium and mercury showed no signs of electrolysis by passing through it an electric current.

Electrolysis of Nitrate of Mercury.—Copper immersed in a solution of nitrate of mercury deposits the latter, and forms an amalgam. By electrolysing a solution of cupric sulphate into a cathode of mercury a similar alloy is formed. An aqueous solution of mercuric nitrate has been used by electro-platers for "quicking" the surfaces of articles previous to plating them. A solution of nitrate of mercury yields its metal to bismuth, zinc, cadmium, lead, iron, or copper, but not to silver, gold, or platinum, by simple immersion.

Electrolysis of Mercuric Chloride.—$Hg.Cl_2$. Moleular weight = 271. From an aqueous solution of this salt magnesium deposits mercuric oxide and calomel (Commaille, *Chem. News*, Vol. XIV., p. 188).

A solution of mercuric chloride, slightly acidulated with sulphuric acid, in a platinum vessel cathode, was electrolysed, and the amount of mercury in it determined by means of a current from six Bunsen cells, the anode being a sheet of platinum. Mercurous chloride was first deposited, but at the end of one hour all the salt was reduced to mercury so perfectly that the supernatant liquid was not rendered cloudy by addition of ammonia. By running a stream of water finally through the vessel whilst the current was passing the whole of the mercury was obtained in a pure state (F. W. Clarke, *Report of the Chemical Society of Berlin*, No. 12, 1878). Gladstone and Tribe noticed that when a weak current was passed through ·a solution of mercuric chloride into a cathode of platinum a film of mercurous chloride was deposited, but if the current was strong, metallic mercury was set free.

Electrolysis of Potassio Mercuric Cyanide.—2KCy.Hg. Cy_2. Molecular weight = 382·2. The solution of this salt readily deposits its metal by simple immersion upon copper, and various of the base and alkali metals, and is therefore used by electro-depositors to prepare, by the process termed "quicking," the surfaces of metal articles to receive an adherent deposit of silver. It also readily yields its metal by means of the other methods of electrolysis.

Electrolytic Movements of Mercury.—In consequence of being a liquid, mercury exhibits certain peculiar phenomena of motion and alteration of form when used as an electrode. This effect appears to be partly a consequence of a film of oxide or other salt formed upon it when it is an anode, and of hydrogen or other substance formed upon it when a cathode, and partly also of simple electrification. H. Herwig has observed that a drop of mercury placed on a glass plate, and strongly electrified by either pole of a Holtz machine, becomes flattened, and if the mercury is in a narrow glass tube its capillary depression is greatly diminished. The effect is greatest with the positive pole, probably in consequence of the higher tension of that pole (*Pogg. Ann.* CLIX, pp. 489—492). As early as the year 1801, Gerboin noticed some of the electrolytic movements of the metal, and the phenomena have since been investigated by Sir H. Davy, Sir J. Herschel, Serullas, Erman, Runge, Poggendorff, Gmelin, and others (*see* Gmelin's "Handbook of Chemistry," Vol. I., pp. 381—384). Also by T. S. Wright (*Phil. Mag.*, Vol. XIX., 1860, pp. 129—133), by R. Sabine (*Phil. Mag.* [5], Vol. II., p. 401), and Th. du Moncel (Watts's "Dictionary of Chemistry," Vol. VIII., Part I., p. 714). The movements have also been applied by Lippmann to the measurement of extremely feeble electric polarities in his capillary electroscope; and also to the production of motion in his capillary electric engine. In most of these cases the electrolyte employed was dilute sulphuric acid.

Electrolytic Sounds.—By employing as an electrolyte a solution of potassio mercuric cyanide, with electrodes of the liquid metal, I discovered that the mercury emitted electrolytic sounds, and became covered on its surface with minute waves, symmetrically disposed, and beautiful in appearance. These waves and sounds appear to be due to the rapid alternate formation and destruction of films upon the mercury by electrolytic action. The best solution for producing it consists of 10 grains of mercuric cyanide and 100 of pure potassic hydrate dissolved in $2\frac{3}{4}$ ounces of aqueous hydrocyanic acid of "Scheele's strength," and the liquid filtered. The waves and sound occur at the cathode. The mercury may be contained in two small watch glasses submerged in the solution contained in a large, flat-bottomed glass basin. The current employed may be from two Grove or five Smee elements, and conveyed into the electrodes by platinum wires protected from the electrolyte by means of glass tubes. By suitable tests I found that during the emission of the sounds the electric current was rendered to a considerable extent intermittent, and that the arrangement might be employed for similar uses to those of a voltaic break-hammer; the intermittence, however, is much less perfect (see *Proc. Roy. Soc.*, 1861 and 1862). For the detection and estimation of mercury by electrolysis, *see* F. W. Clarke, *Jour. Chem. Soc.*, Vol. XXXIV., 1878, p. 916; also Vol. XXXVI., 1879, p. 976; J. Lefort, *ibid.*, Vol. XXXVIII., 1880, p. 510; Watts's "Dictionary of Chemistry," Vol. VIII., Part II., p. 1,277. J. B. Hannay estimates mercury electrolytically by passing a current through a solution of its sulphate into a platinum dish containing it.

Separation of Copper.—Cu. Atomic weight = 63·5. A dyad cation. By simple immersion of magnesium in a solution of cupric chloride, Brunswick green, but no metallic copper, appears; but in one of cupric sulphate, the metal, together with its hydrated protoxide, and a green subsalt are produced (Commaille, *Chem. News*, Vol. XIV., p. 188). From a solution of cupric nitrate or sulphate, aluminium after two days' immersion deposits copper; in the nitrate solution a green basic salt of copper is also produced; but if a minute amount of alkali chloride is added to either of these liquids, deposition commences at once. From a solution of cupric chloride or acetate, aluminium separates copper at once, but the deposition afterwards proceeds slowly (A. Cossa, Watts's "Dictionary of Chemistry," Vol. VII., p. 54). According to Smee iron does not decompose a neutral solution of cupric acetate, nor alkaline ones of ammonuret, ammonio nitrate, or ammonio sulphate of copper, but decomposes one of the nitrate. Zinc amalgam deposits copper from neutral solutions of cupric salts, and forms a copper amalgam (Damour, *Jour. Prac. Chem.*, Vol. XVII., p. 345). By adding crystals of silicon to melted protoxide of copper, I observed that sudden incandescence and a full white heat were produced, and metallic copper was separated. Thallium deposits copper from the solution of cupric nitrate, sulphate, and acetate (W. C. Reid, *Chem. News*, Vol. XII., p. 242). Raoult states that gold in contact with copper, in either a cold or boiling acid or neutral solution of a cupric salt, receives no deposit of copper (*Jour. Chem. Soc.*, Vol. XI., p. 465).

Smee states that the use of solutions of the hyposulphite, ammoniuret, or acetate of copper, with a separate current, offers no advantages for depositing copper, because they are difficult to decompose, and require a current from several cells, that a copper anode is but little corroded in a solution of sulphocyanide of potassium, and the solution does not hold much dissolved metal; also that the anode is very slightly acted upon in a solution of tartrate of potassium. M. P. Schutzenberger found from five to ten per cent. of cuprous oxide in copper electro-deposited from an acetate solution (J. B. Mackintosh, *Chem. News*, Vol. XLIV., p. 279, and Vol. XLV., p. 101).

Formation of Nitride of Copper.—By passing a current from six Grove cells into one end of a solution of sal ammoniac contained in a long glass trough by means of a copper anode, and out of the liquid at the distant end by a platinum sheet cathode, the liquid becomes blue, and a heavy solid nitride of copper of a chocolate colour collects at the cathode (Grove, *Phil. Mag.*, 3rd Series, Vol. XIX., p. 100).

Electrolysis of Cupric Nitrate.—$Cu.2No_3$. Molecular weight $= 187\cdot5$. I observed that a solution of cupric nitrate yielded its metal to zinc, tin, lead, or iron, by simple immersion, but not to nickel, copper, silver, gold, platinum, or antimony. J. B. Mackintosh states that in the electro-deposition of copper from a nitrate or sulphate solution containing citric or tartaric acid, the deposited metal is not pure. With the nitrate solution containing citric acid, electrolysis was attended by a strong odour of hydrocyanic acid (*Chem. News*, Vol. XLIV., p. 279, and Vol. XLV., p. 101).

Electrolysis of Cupric Fluoride.—$Cu.F_2$. Molecular weight $= 101\cdot5$. I observed that copper was separated from its melted fluoride by adding fragments of magnesium ; also that crystals of silicon immersed in a solution of the fluoride evolved gas, and became instantly coated with copper.

By means of a separate current from six Smee cells, a platinum wire helix anode and a copper wire helix cathode, I electrolysed fluoride of copper, fused at a bright red heat in a deep copper cup. Conduction was copious, as if the salt was a metal, and an acid vapour was evolved. The anode was unaltered, no copper was deposited, but the cathode had lost $3\cdot35$ grains in weight by corrosion near the surface of the melted salt ; the copper vessel was also similarly acted upon in several experiments and caused to leak. The phenomena were much like those obtained with melted argentic fluoride.

A solution of cupric fluoride in pure dilute hydrofluoric acid, with copper electrodes, conducted freely the current from a single Smee element, and yielded a good deposit of copper.

Electrolysis of Cupric Chloride. — $Cu.Cl_2$. Molecular weight $= 134\cdot4$. Aluminium acts briskly on a solution of cupric chloride at 16°C, setting free copper, hydrogen, and aluminium oxychloride, the composition of which varies with the temperature (D. Tommasi, *Jour. Chem. Soc.*, Vol. XLII., 1882, p. 1,266; Vol. XLIV., 1883, p. 19; *Chem. News*, Vol. XLVI., p. 62). Copper is at once deposited from its chloride, and more slowly from its acetate, by aluminium (A. Cossa, Watts's "Dictionary of Chemistry," Vol. VII., pp. 54 and 383). I observed that in a solution of cupric chloride, bismuth, zinc, tin, lead, and iron deposited copper by simple immersion, but nickel, copper, silver, gold, platinum, or antimony did not; also that in a solution of sub-chloride of copper in strong aqueous ammonia zinc received a deposit of copper by simple immersion, but tin, lead, iron, nickel, copper, silver, gold, platinum, bismuth, or antimony did not. When a copper platinum couple is immersed in a dilute solution of this salt insoluble white cupreous chloride is deposited on both the metals. With couples formed of zinc-platinum or magnesium-platinum the action is stronger, and metallic copper is deposited upon the platinum (Gladstone and Tribe, *Phil. Mag.*, 4th Series, Vol. XLIX., p. 425).

M. Weis Kopp coats iron with copper by simply immersing it in a bath composed of 10 parts of cupric chloride, 10 of nitric acid, and 80 of hydrochloric acid of specific gravity 1·105 (*Chem. News*, Vol. XXL, p. 47). According to O. Gaudain, articles of cast iron, wrought iron, or steel may be coated with copper by dipping them into a melted mixture of fluoride and chloride of copper, with five or six parts of cryolite, and a little basic chloride, in a plumbago crucible (*Jour. Chem. Soc.*, Vol. XI., p. 955). In some cases, copper is extracted from sandstone, which contains it in small proportion, by dissolving the ore out by dilute hydrochloric acid, and immersing pieces of scrap iron in the liquid until they are wholly dissolved, and metallic copper is left.

When a feeble electric current is passed by means of copper electrodes through a solution of chloride of copper in dilute hydrochloric acid, the anode becomes covered with snow-white crystals of cupreous chloride, and the cathode with a thick deposit of loose copper (*Chem. News*, Vol. XXII., p. 167). If a solution of cupric chloride was electrolysed by a feeble current with platinum electrodes, chlorine appeared at the anode and cupreous chloride at the cathode; but if the current was strong, metallic copper was also deposited upon the edges of the cathode (Gladstone and Tribe). Smee states that a solution of this salt is less readily decomposed by an electric current than one of the nitrate, but more readily than one of the sulphate, that it is also one of the worst liquids for the electrolytic separation of metallic copper, and that the deposited metal is apt to assume a very peculiar appearance. He also states that a solution of the ammonio-chloride is a bad one, having a tendency to evolve hydrogen and yield a spongy copper deposit ; and that one composed of iodide of copper dissolved in aqueous solution of potassic iodide cannot be employed because it liberates iodine. The electrolysis of solution of cupric bromide does not appear to have been examined.

Electrolysis of Cupric Sulphate.— $Cu.SO_4$. Molecular weight $= 127·5·$ I have noticed that a solution of cupric sulphate gave up its metal by simple immersion of zinc, tin, lead, or iron, but not to nickel, copper, silver, gold, platinum, bismuth, or antimony. One of the oldest facts in electro-chemistry is the deposition of copper upon iron by immersing the latter in an aqueous solution of cupric sulphate. Vast numbers of steel pens, iron wire, and other small articles of steel and iron are coated with copper by means of the same liquid slightly acidulated. For coating iron wire in this way Roseleur used a mixture composed of one part of cupric sulphate and one of sulphuric acid dissolved in from 50 to 100 parts of water. To coat brass with copper, Dr. C. Puscher dissolves 10 parts of cupric sulphate and 5 of ammonic chloride in 150 of water ; dips the clean articles in the liquid for one minute, drains them, and then heats them over a charcoal fire until the ammoniacal salt is expelled, and then washes and dries them (*Chem. News*, Vol. XXIII., p. 215). According to M. Soret, clean copper dissolves in a boiling, saturated, aqueous solution of neutral cupric sulphate, and is deposited in a metallic state on cooling the liquid (*Annal. de Chimie*, Vol. XLII., 1854, pp. 257—277). I have found only a small quantity of red suboxide of copper separate under these conditions. According to Wurtz, pure cupreous hydride can be obtained by the electrolysis of a dilute solution of copper sulphate (*Jour. Chem. Soc.*, Vol. XXXVIII., 1880, p. 299).

Cupric sulphate resulting from the oxidation of cupric sulphide in the earth exists in the water of many mines. Immense quantities of such sulphide are also roasted with common salt to oxidise and render soluble the cupriferous mineral, which is then dissolved out by dilute hydrochloric acid, and the copper in each of these cases is extracted from the liquid by immersing in it scraps of iron. The copper collects as a red powder consisting of small feathery crystals at the bottom of the vats after all the iron has dissolved. Great quantities of the metal are annually deposited in this manner.

Copper is sometimes deposited upon iron by the influence of the contact of a second metal. M. Fred. Wiels uses the following liquid :—Dissolve 150 parts of sodio-potassic tartrate, 80 of soda lime containing from 50 to 60 per cent. of caustic soda, and 35 of cupric sulphate, in 1,000 parts of water. By immersing clean articles of iron or steel in contact with a piece of zinc or lead in this liquid a sufficiently long period of time they receive a strongly adherent coating of copper, of any desired thickness. Pure tin in contact with zinc in this liquid does not become coppered, but oxidises, and its oxide gradually precipitates as red suboxide all the copper of the solution.

Copper is also deposited by means of the contact of two different metals in two different liquids (the " single cell process ") for the purpose of coating cast-iron cylinders for calico printing (see *Chem. News*, Vol. XXX., p. 219 ; also *Jour. Chem. Soc.*, Vol. XIII., p. 196).

A good solution for depositing by means of a separate current is composed of four parts of crystallised cupric sulphate ($Cu.SO_4, 5H_2O$) and one of sulphuric acid, in 18 or 20 of water. Sometimes sulphate of zinc or of potassium is added to such a solution in order to improve the deposit.

According to A. Long, copper electro-deposited by a separate current from a solution of its sulphate contains minute amounts of hydrogen, carbonic oxide, carbonic anhydride, and water (Watts's " Dictionary of Chemistry," Vol. VII., p. 383).

Elimination of Impurities from Copper by means of Electrolysis.—Very few metals are likely to be electro-deposited along with copper from the usual acid sulphate depositing solution by a separate current ; the most likely one is cadmium. In the electrolysis of that solution with an anode of ordinary copper, a considerable amount of black insoluble matter separates at the anode. An analysis of this substance by Max Duke of Leuchtenberg gave the following :—

Tin	33·50
Oxygen	24·82
Copper	9·24
Antimony	9·22
Arsenic	7·20
Silver	4·45
Sulphur	2·46
Nickel	2·26
Silicia	1·90
Selenium	1·27
Gold	·98
Cobalt	·86
Vanadium	·64
Platinum	·44
Iron	·30
Lead	·15 = 99·69

(Erdmann, *Jour. Prac. Chem.*, Vol. XLV., pp. 460—468.)

The following are comparatively recent analyses kindly supplied to me by a friend. They are those of the powder from

copper plates used as anodes in depositing copper on statues,
&c. :—

No. I.		No. II.		No. III.	
Copper	85·550	Lead...............	27·70	Copper	67·90
Water and oxy-		Water and oxy-		Sulphur	18·10
gen	4·950	gen	21·05	Iron	5·55
Arsenic	2·480	Copper............	19·40	Insoluble earthy	
Silver	1·815	Antimony	7·35	matter	3·40
Sulphuric acid..	1·150	Sulphur	6·35	Organic matter..	2·25
Insoluble earthy		Silver	5·61	Lead	2·05
matter.........	·950	Arsenic	5·20	Silver	·55
Antimony	·750	Earthy matter...	4·35	Loss	·20
Iron...............	·750	Bismuth	1·25		
Bismuth	·650	Chlorine	·70		
Alumina	·250	Iron	·60		
Chlorine	·250	Nickel	·20		
Gold	·085	Organic matter..	·20		
Lead	·050	Gold...............	·01		
Loss	·020	Loss	·03		
	100·000		100·00		100·00

Refining of Crude Copper by Electrolysis.—This process is carried out on a large scale by Messrs. Elkington, at
their copper works, Pembrey, South Wales. The process
simply consists in making large slabs of the crude metal,
obtained by the ordinary smelting process, anodes in the usual
cupric sulphate solution, and passing currents from numerous
dynamo electric machines through the solutions, until the
slabs are wholly dissolved and their copper deposited. Each
current passes in an undivided state through a series of such
electrodes and solutions in order to diminish the cost of the
process.

The impurities which separate vary, of course, with the
different samples of crude metal. In the process, oxygen,
sulphur, selenium, carbon, boron, silicon, and arsenic are not
deposited at the cathode. Silver is precipitated at the anode
by the traces of hydrochloric acid present in the common
sulphuric acid employed. Gold falls as metal at the anode,
lead as sulphate ; carbon and metallic sulphides, also selenium
and silica, fall at the anode. Zinc, iron, tin, cadmium, cobalt,
nickel, and antimony are more or less dissolved, but, being
less readily deposited than copper, remain in solution. Arsenic
falls as an arsenide, and the metal most likely to be deposited,
viz., cadmium, is present so rarely, or in so small an amount,
as to remain in solution.

Electric Etching, &c.—Copper being a very suitable metal
for the purposes of engravers and printers, the electric corrosion of anodes in a solution of cupric sulphate was soon applied
to engraving and etching. Copper being also a metal easily
deposited, the process of electro-deposition of copper in solutions of its sulphate was also applied to the multiplication of

G 2

engraved plates, the copying of set-up type, the manufacture of works of art, and even of colossal statues, &c. The details of these and other technical uses of electro-chemical action may be found described in works on Electro-Metallurgy.

For the uses of electrolysis in the metallurgy of copper by the processes of Becquerel, Keith, and Patera, see *Jour. Chem. Soc.*, Vol. XXXVI, 1879, p. 760 ; also Blas and Miest, *Chem. News*, Vol. XLVI, p. 121. The latter also apply electrolysis to all kinds of ores.

Electrolysis of Cuproso Potassic Cyanide.—There are several cuproso cyanides of potassium. The one usually employed for electro-deposition is formed by dissolving green cuproso cupric cyanide to the point of saturation in a solution of potassic cyanide, and then adding some more of the potassic solution, and using the mixture at a temperature of about 150°F. The base metals much less readily deposit copper by simple immersion in this liquid than in the ordinary cupric sulphate solution. By the passage of a separate current this salt is also less readily decomposed than cupric sulphate, and hydrogen is freely set free at the cathode along with the copper.

Various mixtures of salts containing potassic cyanide have been employed for depositing copper upon base metals. Roseleur recommends the following :—Rub 20 parts of crystallised verdigris to powder in a little water, add to it with stirring 20 parts of washing soda dissolved in 200 parts of water, mix the solution with one of 20 parts of bi-sulphite of sodium dissolved in 200 parts of water, and add the mixture with stirring to a solution of 20 parts of pure potassic cyanide dissolved in 600 parts of water ; then if the mixture is not colourless, add more potassic cyanide until it is so. It may be used either hot or cold. A second one he recommends is composed of 20 parts of strong aqueous ammonia, 30 of sodic bi-sulphite, 35 of cupric acetate, 50 of potassic cyanide of 70 per cent., and 2,500 of water. The ammonia and copper salt are dissolved in one portion of the water, and the cyanide and bi-sulphite in the other, and the two solutions mixed. If the resulting solution is at all blue, more potassic cyanide must be added to render it colourless. This mixture also may be used hot or cold.

Another liquid employed for the same purpose may be made by dissolving 40 parts of the blue ammoniuret of copper and 80 parts of potassic cyanide in about 1,000 parts of water. W. H. Walenn recommends cyanide of copper dissolved to saturation in an aqueous solution of equal parts of potassic cyanide and ammonium tartrate, and sufficient oxide and ammoniuret of copper added to prevent evolution of hydrogen at the cathode when the liquid is used at 80°C. with a current from a single Smee element. Dr. Elsner used a solution composed of one part of potassic bitartrate boiled in ten parts of water, and as much freshly-prepared and wet hydrated cupric carbonate which has been washed with *cold* water, stirred with it, as the liquid will dissolve. A small quantity of potassic carbonate is then added. He states that a copper anode dissolves readily in this mixture.

According to F. Weil, by employing an alkaline solution in which cyanides are replaced by organic acids or glycerol, copper may be firmly deposited by a separate current on wrought iron, cast iron, and steel, and the acids or glycerol are not decomposed (*Jour. Chem. Soc.*, Vol. XLII., 1882, p. 670).

Copper has been deposited upon iron by the combined action of simple immersion and of a separate current in a solution of one part of cupric oxalate and a large excess of potassic bi- or quad-oxalate in ten to fifteen parts of water (Watts's "Dictionary of Chemistry," Vol. VIII., Part II., p. 1,118).

The physical properties of the copper deposited from the various mixtures, and from each solution at different temperatures, or by different strengths of current, vary considerably. A trace of carbonic bisulphide in the cupric sulphate solution makes the deposit brittle, the anode also becomes black, but if there is also a great excess of acid, it sometimes becomes very bright ; and if the liquid also contains much potassic sulphate, the deposited copper is said to be bright. The deposit also from the cyanide is usually bright when the current is strong, and of a dull aspect when it is weak. According to Favre, electro-deposited copper contains more heat than the rolled metal (Watts's "Dictionary of Chemistry," Vol. VII., p. 462). For the absorption of gases by deposited copper, *see* Watts's "Dictionary of Chemistry," Vol. VII., p. 383.

Analysis of Copper Ores by Means of Electrolysis.—As this series of articles is not of a technical character very few remarks are admissible on this subject. To carry out also pro-cesses of electrolytic analysis successfully, requires a knowledge of analytical chemistry, because the methods in nearly all cases (with other metals as well as with copper) are combinations of ordinary chemical and electro-chemical actions.

By electrolysis all the copper is separated from solutions containing free hydrochloric acid on the addition of ammonium or sodium chlorides, or sodium acetate ; similarly from solutions containing excess of ammonia, ammonium carbonate, or potassic cyanide. From a solution containing mercury, silver, bismuth, and copper, the last two metals are only deposited after the greater portion of the first two has separated (C. Luckow, *Jour. Chem. Soc.*, Vol. XXXVIII., 1880, p. 283).

The electrolytic determination of the amount of copper pre-sent in a liquid is more readily made than that of almost any other metal, and this agrees with the usually extreme degree of purity of the deposited substance. The last traces of copper may also be perfectly precipitated in a coherent state from a solution of blue vitriol containing two platinum elec-trodes, by a current of suitable strength. The deposited copper, however, is not perfectly pure if tartaric or citric acid is present. The electrolytic process is exten-sively used. Details of it may be found in Watts's "Dic-tionary of Chemistry," Vol. VII., pp. 384, 790; Vol. VIII., p. 559: *Chem. News*, Vol. XIX., 1869, p. 221; XXIV., pp. 100 and 172; XLI., pp. 25, 213; XLII., p. 331; XLIV., 1881, p. 279; XLV., 1882, p. 101, and XLVI., p. 105 : *Jour. Chem. Soc.*, 1876, Part II., p. 115; 1877, Part I., p. 340; Vol. XXXVI., 1879, p. 276; Vol. XXXVIII., 1880, pp. 282 and 583; Vol. XL., 1881, p. 1,081 ; Vol. XLII., 1882, pp. 428, 660, 896, and 1,320.

Separation of Nickel.—Ni. Atomic weight = 59. A dyad cation. Less readily deposited than copper. From slightly acid solution of salts of protoxide of nickel, magnesium de-posits by simple immersion metallic nickel and hydrogen (Roussin, *Chem. News*, Vol. XIV., p. 27). Zinc amalgam deposits nickel from neutral solutions of nickel salts by simple immersion, and forms an amalgam (Damour, *Jour. Prac. Chem.*, XVII., p. 345).

By contact with a second metal, nickel is also in some cases deposited from its solutions. Stolba takes a boiling hot, one-third saturated solution of chloride of zinc, in a copper vessel, renders it clear by adding just sufficient hydrochloric acid, then adds a few particles of zinc, sufficient to cause a slight deposit of zinc upon the copper. He next adds either chloride or sulphate of nickel, until the mixture is distinctly green. The metals to be coated, viz., cast iron, wrought iron, steel, brass, or copper, are then immersed in the boiling solu-tion in contact with zinc until they are coated (Watts's "Dic-tionary of Chemistry," Vol. VII., p. 850 ; also *Chem. News*, Vol. XXXV., p. 166). C. Mène coats either iron, steel, zinc, lead, copper, or brass with nickel, by immersing it in a boil-ing hot neutral solution of chloride of zinc, containing frag-ments of nickel. If the liquid is acid, the deposit appears dull (*Chem. News*, Vol. XXV., p. 214). A nickel-gold couple produces no deposit of nickel in acid or neutral, hot or cold, solu-tions of salts of nickel (Raoult, *Jour. Chem. Soc.*, Vol. XL, p. 646).

An aqueous solution of cream of tartar and hydrated nickel oxide, with a little soda, gave by the separate current process peroxide of nickel at the anode (Wernicke, Watts's "Dictionary of Chemistry," Vol. VII., p. 899).

Electrolysis of Nitrate of Nickel.—Ni.2NO$_3$. Molecular weight $= 183$. Nickel may be deposited by a separate current from a solution formed by dissolving one part of nitrate of nickel in one part of strong aqueous ammonia, and then adding 20 to 30 times its volume of aqueous bisulphate of sodium of specific gravity 1·999 (Roseleur). I have always found that when nickel solutions contained nitrates the deposited metal was of a bad colour.

. In France a solution is prepared by dissolving four parts of nickel nitrate in four parts of aqueous ammonia and 150 parts of water holding in solution 50 parts of acid sodium sulphite. A very feeble current is used (Boden, Watts's " Dictionary of Chemistry," Vol. VIII., Part II., p. 1,388).

Electrolysis of Fluoride of Nickel.—By immersing crystals of silicon in an aqueous solution of nickel fluoride, containing free hydrofluoric acid, I observed that they did not become coated with metal; but by heating the crystals with ten times their weight of nickel fluoride to redness in a porcelain crucible, vivid incandescence occurred, and nickel was deposited and melted by the great heat evolved.

Electrolysis of Chloride of Nickel.—Ni.Cl$_2$. Molecular weight $= 130$. The simple immersion of copper in a solution of the double chloride of nickel and sodium is sufficient to deposit the nickel (Becquerel, *The Chemist*, Vol. V., p. 408). Zinc throws down the metal from a solution of nickel chloride previously rendered alkaline by addition of ammonia.

. One of the first really good liquids for depositing nickel by means of a separate current for practical purposes appears to have had its origin in the following experiments published by me :—
" I have deposited nickel in the state of reguline white metal from a solution of the double chloride of nickel and ammonium, by making a lump of metallic nickel the anode in a strong aqueous solution of hydrochlorate of ammonium (sal ammoniac), and passing a strong current until the liquid acquired a pale greenish-blue colour" (*Pharm. Jour.*, Vol. XV., No. 9, September 1, 1855, pp. 106 and 131). T. Fearn, in 1872, published the composition of a solution for depositing nickel, viz., 24 parts of sal ammoniac dissolved in 160 parts of water, and the liquid then saturated with protoxide of nickel at 120°F.

Martin and Dalmotte dissolve 1,250 grammes of citric acid, 500 of sal ammoniac (or ammonium sulphate) and 500 of nitrate of ammonium in 15 litres of water; heat the liquid to 80°C., and saturate it with recently precipitated hydrate of nickel, then add 2½ litres of aqueous ammonia; dilute to 25 litres, and after cooling add 500 grammes of ammonic car-.

bonate, subside the mixture, and filter the liquid. Use the solution at 50°C. (Watts's "Dictionary of Chemistry," Vol. VIII., Part II., p. 1,388).

Electrolysis of Sulphate of Nickel.—$Ni.SO_4$. Molecular weight = 155. Magnesium deposits nickel by simple immersion from a solution of nickel sulphate (Commaille, *Chem. News*, Vol. XIV., p. 188). Zinc throws down the metal perfectly from a solution of nickel sulphate rendered alkaline by addition of ammonia (A. Merry, *Jour. Chem. Soc.*, Vol. XIII., p. 311).

The best solution for electro-depositing nickel is made either by dissolving the crystallised double sulphate of nickel and ammonium, in the proportion of half a pound to a pound, in a gallon of water, or the double chloride of nickel and ammonium may be used instead. A large anode of nickel should be employed. Böttger states that the best solution for depositing nickel by means of a separate current is made by adding to crystals of proto-sulphate of nickel as much liquid ammonia as is necessary to dissolve them (*Pharm. Jour.*, Vol. III., 1843, p. 358). Nagel dissolves two parts by weight of crystals of sulphate of nickel in a mixture of six parts of aqueous ammonia of specific gravity ·909 and thirty parts of water, and uses the mixture at a temperature of about 100°F. Another liquid is composed of 100 parts of sulphate of nickel, 53 of tartaric acid, and 14 of hydrate of potassium, dissolved in a suitable proportion of water. It is said to yield a bright deposit of metal. Some recipes include nitrate of ammonium, or nitric acid, which is objectionable (*see* also A. C. and E. Becquerel, *Comptes Rendus*, Vol. LV., p. 18 ; also Kayser, *Jour. Chem. Soc.*, Vol. XXXIV., 1878, p. 537).

Another nickel solution is composed of 87·5 parts of nickel sulphate, 20 of ammonium sulphate, 17·5 of citric acid, and 2 litres of water (Hesse, Watts's "Dictionary of Chemistry," Vol. VIII., Part II., pp. 1,118 and 1,388).

More recently nickel-sulphate-depositing solutions containing borax have been employed. I analysed one, and found it to contain sulphate and chloride of nickel, borax, and a small quantity of ammonia. A Mr. Powell, of Cincinnati, adds 1⅛oz. to 1oz. of benzoic or pyrogallic acid to each gallon of the ordinary nickel plating solution " to improve it."

A solution of ferrocyanide of nickel dissolved in aqueous potassic cyanide has also been employed for depositing the metal.

In the deposition of nickel from the solution of the double sulphate of nickel and ammonium with a cast nickel anode the anode disintegrates to a loose powder upon its surface, and also by solution of the nickel a loose coating of impurity accumulates upon it, and falls to the bottom of the liquid and collects as a yellow mud. The following are the results of a chemical analysis of the yellow substance :—

Hydrated oxide of nickel	52 7
SO₃ (chiefly insoluble basic sulphates)	11·5
Moisture	11·3
Sesquioxide of iron	9·4
Sand (with particles of graphite)	8·6
Metallic nickel	4·8
Oxide of copper	·7
	99·0

Also a trace of sulphur, but no silver, lead, or zinc, nor any metallic iron.

Electrolysis of Selenate of Nickel.—By adding to a solution of neutral selenate of nickel aqueous ammonia until the liquid was of a clear blue colour, and electrolysing with a nickel anode and a current from three Smee cells, I obtained a brilliant and very white deposit of the metal.

Electrolytic Analysis of Nickel Compounds.—See *Chem. News*, Vol. XXIV., pp. 100 and 172; Vol. XXVI., p. 209; Vol. XXXVIII., p. 26; Vol. XLI., p. 25; Vol. XLII., pp. 75 and 331; Vol. XLVI. p. 105. *Jour. Chem. Soc.*, 1876, Part II., p. 115; Vol. XL, p. 204, Part I., 1877, p. 340, and Part II., pp. 924 and 925; Vol. XXXIV., 1878, p. 537; Vol. XXXVIII., 1880, pp. 284, 583, 751, and 771; Vol. XL., 1881, p. 1,081; Vol. XLII., pp. 896, 1,320; Vol. XLVI., p. 105. Watts's "Dictionary of Chemistry," Vol. VII., pp. 791, 849.

Separation of Cobalt.—Co. Electro-chemical equivalent $\frac{59}{2} = 29·5$. A dyad cation. Magnesium deposits metallic cobalt and hydrogen from slightly acid solutions of protoxide of cobalt (Roussin, *Chem. News*, Vol. XIV., p. 27). Thallium deposits a basic salt by simple immersion in a solution of nitrate of cobalt (W. C. Reid, *Chem. News*, Vol. XII., p. 242). Zinc amalgam deposits cobalt by simple immersion in neutral solutions of salts of cobalt, and forms an amalgam (Damour, *Jour. Prac. Chem.*, XVII., p. 345). Cobalt is not precipitated from its neutral solutions by means of zinc, except in the presence of a metal easily reducible by zinc, *e.g.*, lead or copper, but not cadmium; with copper salt present, if the liquid is acid, copper alone is deposited. A definite quantity of copper salt is necessary (Lecoq de Boisbaudran, *Jour. Chem. Soc.*, 1876, Part II., p. 551). Cobalt is deposited upon steel or iron by contact of zinc in a boiling hot solution of zinc chloride containing a salt of cobalt (*Chem. News*, Vol. XXXV., p. 166). Mène deposits cobalt upon lead, iron, brass, or copper by immersing it in contact with zinc in a boiling hot and neutral solution of chloride of zinc containing fragments of cobalt (*Chem. News*, Vol. XXV., p. 214).

Formation of Peroxide of Cobalt.—An aqueous solution of cream of tartar and hydrated cobalt oxide, with a little soda dissolved in it, yields, with a separate current, a peroxide of cobalt, exhibiting magnificent colours upon the anode (W. Wernicke, Watts's "Dictionary of Chemistry," Vol. VII., p. 899).

By passing a separate current through a solution of oxide of cobalt in aqueous potassic cyanide, hydrogen and a small quantity of cobalt are deposited.

According to Troost and Hautefeuille, laminæ of electro-deposited cobalt sometimes contain as much as thirty-five times their volume of hydrogen (*Chem. News*, Vol. XXXI., p. 196).

Electrolysis of Fluoride of Cobalt.—$Co.F_2$. Molecular weight $= 97$. I electrolysed a solution of this salt in pure dilute hydrofluoric acid, by means of a current from a single Smee cell, with an anode of cobalt and a cathode of copper, but only a film of black powder appeared on the cathode in twelve hours.

Electrolysis of Chloride of Cobalt.—$Co.Cl_2$. Molecular weight $= 130$. Magnesium decomposes a solution of cobalt chloride, with evolution of hydrogen and separation of a green salt containing cobalt oxide (S. Kern, *Jour. Chem. Soc.*, 1876, Part I., p. 880). Copper immersed in a solution of the double chloride of cobalt and sodium acquires a coating of cobalt (Becquerel, *The Chemist*, Vol. V., p. 408).

In a solution composed of 20 parts of sal ammoniac, 40 of chloride of cobalt, 20 of aqueous ammonia, and 100 of water, a brilliant deposit of metallic cobalt was produced upon a cathode of brass or copper, by means of a current from two Bunsen cells (M. R. Boettger, *Chem. News*, Vol. XXXV., p. 166, also *Jour. Chem. Soc.*, 1877, Part II., p. 375. To deposit the metal, dissolve five ounces of its dry chloride in a gallon of distilled water, and make the solution slightly alkaline by means of aqueous ammonia. Pass a current from three to five Smee cells through the solution by means of an anode of cobalt (*Telegraphic Journal*, Vol. II., p. 246). By means of a separate current, an anode of cobalt, and a concentrated solution of the chloride, with its excess of acid neutralised by caustic ammonia, Becquerel obtained deposits of the metal, brilliant, white, hard, and brittle, and possessing magnetic polarity. He observed that part of the chloride of the solution was set free during the electrolysis, and that if the liquid contained iron the greater portion of it was not deposited with the cobalt (*Chem. News*, Vol. VI., p. 126).

Electrolysis of Sulphate of Cobalt.—$Co.SO_4$. Molecular weight $= 155$. Magnesium slowly deposits hydrated oxide of cobalt from a solution of the sulphate (Commaille, *Chem. News,* Vol. XIV., p. 188).

By means of a separate current, cobalt is completely precipitated in the metallic state from an aqueous solution of double sulphate of cobalt and ammonium, if free ammonia is present (H. Fresenius and F. Bergmann, *Chem. News,*Vol. XLII., p. 75). Gaiffe deposited hard tenacious cobalt of good colour from an aqueous solution of the double sulphate of cobalt and ammonium by means of a separate current (*Chem. News,* Vol. XL., p. 23).

Electrolytic Analysis of Compounds of Cobalt.—See *Chem. News,* Vol. XLI., p. 25 ; Vol. XLII., p. 75 ; Vol. XLVI., p. 105 ; *Jour. Chem. Soc.,* 1877, Part I.. p. 341 ; 1877, Part II., p. 925 ; Vol. XXXVI., 1879, p. 588 ; Vol. XXXVIII., pp. 284, 583, and 771 ; Vol. XL., 1881, p. 1,081 ; Vol. XLII., 1882, pp. 896, 1,320.

Separation of Iron. — Fe. Electro-chemical equivalent $\frac{56}{2} = 28$. A dyad cation. From slightly acidified solutions of ferrous and ferric salts, magnesium deposits iron and hydrogen gas (Roussin, *Chem. News,* Vol. XIV., p. 27). Iron in contact with gold, in acid or neutral, cold or hot solutions of salts of iron, produces no metallic deposit (Raoult, *Jour. Chem. Soc.,* Vol. XI., p. 646). Metallic iron reduces ferric to ferrous salts at ordinary temperatures, whilst platinum has no such effect. Nevertheless, if these two metals are connected together, they reduce the ferric salt more rapidly than iron alone does, and the reduced salt forms upon the platinum also, as may be seen by mixing a little ferricyanide of potassium with the liquid (Gladstone and Tribe, *Phil. Mag.* [4], Vol. XLIX., p. 425).

By electrolysis with a separate current iron is incompletely deposited as metal from neutral solutions of ferrous salts, some ferric salt being formed. If to the neutral solution of ferrous sulphate some ammonium citrate be added containing free citric acid, and care be taken that free citric acid remains in the solution, the iron will be deposited in the metallic lustrous form. No iron is separated by electrolysis from a solution of ferrocyanide of potassium, but prussian blue appears at the cathode. From the solutions of ferrous oxide in solution of sodium thio-sulphate, all the iron is separated, chiefly as ferrous sulphide. From the fluoride of iron dissolved in a solution of sodium fluoride, metallic iron is deposited (C. Luckow, *Jour. Chem. Soc.,* Vol. XXXVIII., 1880, p. 284).

Electrolysis of Ferrous Chloride.—$Fe.Cl_2$. Molecular weight $= 127$. According to Aikin, iron amalgam is formed by the action of zinc amalgam on ferrous chloride; but according to Damour it cannot be produced in this way (Watts's "Dictionary of Chemistry," Vol. III., p. 888). When zinc amalgam is immersed in a solution of ferrous chloride, and a crystal of a nitrate is placed upon it, a black spot is gradually formed upon the surface of the amalgam, consisting of reduced iron, which is immediately taken up by the mercury. Chlorates and other salts do not produce it (Runge, Watts's "Dictionary of Chemistry," Vol. III., p. 891).

The aqueous solution of ferrous chloride yields by electrolysis chlorine and oxygen at the anode, and iron and hydrogen at the cathode (Watts's "Dictionary of Chemistry," Vol. III., p. 377).

Electrolysis of Ferric Chloride.—Fe_2Cl_6. Molecular weight = 325. Ferric chloride is partly reduced to ferrous chloride, partly to metallic iron, by contact with sodium amalgam and a little water ; and by contact with a sufficient quantity of the amalgam it is reduced to metal, which remains as iron amalgam (Cailletet, Watts's "Dictionary of Chemistry," Vol. VI., p. 816).

With magnesium and platinum in contact with each other in a solution of ferric chloride, metallic iron is soon deposited on the platinum. The passage of a feeble current, by means of platinum electrodes through a similar solution, sets free chlorine at the anode and ferrous chloride at the cathode, but a stronger one deposits metallic iron upon the cathode (Gladstone and Tribe, *Phil. Mag.*, 4th Series, Vol. XLIX., p. 425).

E. Becquerel found that when sesquichloride of iron, Fe_2Cl_3, was electrolysed, one atom of chlorine and $\frac{2}{3}$ atom of iron are separated for each atom of hydrogen in the voltameter (Watts's "Dictionary of Chemistry," Vol. II., p. 439). A concentrated acid solution of ferric chloride yields by electrolysis chlorine and a small quantity of oxygen at the anode, and ferrous chloride at the cathode (Watt's "Dictionary of Chemistry, Vol. III., p. 378).

I have deposited metallic iron in a reguline state by passing a current from 15 or 20 Smee cells through a solution of sal ammoniac, by means of an anode of sheet iron and a cathode of copper, for some time, until sufficient iron had dissolved. M. Cailletet states that by electrolysing a solution of ferrous chloride mixed with sal ammoniac the iron was deposited in the form of mammillary masses, brittle, brilliant, and hard enough to scratch glass, and the deposit, when plunged into water, evolved numerous bubbles of pure hydrogen. Also that one volume of the iron absorbed about 240 volumes of hydrogen, which ignited by contact with a flame and surrounded the metal with a pale colour (*Chem. News*, Vol. XXXI., p. 119 ; also *Jour. Chem. Soc.*, Vol. XIII., p. 425). According to R. Leng, deposited iron contains 185 times its volume of hydrogen, chiefly in the layers of metal first deposited (*Chem. News*, Vol. XXI., p. 179), and Troost and Hautefeuille say that it sometimes contains as much as 260 times its bulk (*Chem. News*, Vol. XXXI., p. 196).

Electrolysis of Ferrous Sulphate.—$Fe.SO_4$. Molecular weight $= 152$. Sodium amalgam decomposes a solution of ferrous sulphate, and produces an amalgam of iron (Böttger, Watts's "Dictionary of Chemistry," Vol. III., p. 887). A similar amalgam is formed by electrolysing the same solution with a cathode of mercury (Joule, *ibid.*, p. 888). Magnesium deposits from a neutral solution of ferrous sulphate hydrated ferrous oxide ; but from an acidified one it deposits metallic iron (Commaille, *Chem. News*, Vol. XIV., p. 188). According to Fischer, zinc immersed in a perfectly neutral solution of ferrous sulphate, contained in a stoppered bottle, throws down metallic iron, partly on the zinc. I have found with this solution that neither antimony, bismuth, tin, lead, iron, nickel, copper, brass, German silver, silver, gold, or platinum received a metallic deposit by simple immersion.

I have deposited firm reguline iron, by means of a separate current, from a saturated aqueous solution of a mixture of two parts of ferrous sulphate and one of sal ammoniac. Walenn deposited reguline, white, silvery-looking iron, together with much hydrogen gas, from a cold and slightly acid solution composed of one part of crystallised ferrous sulphate and five of water, by means of a current from three Smee elements of very large surface. The addition of sulphate of ammonium increased the conducting power, and formed a very good conducting liquid (*Chem. News*, Vol. XVII., p. 170). Klein employed a solution of ferrous sulphate as pure, neutral, and cencentrated as possible ; also a feeble electric current. These conditions are important. The iron then obtained was as hard as tempered steel, and very brittle ; but after annealing it was malleable, and might be engraved as easily as soft steel. It had a specific gravity of 8·139, and contained occluded in it 13 times its volume of hydrogen. It possessed a higher electric conductivity than any commercial iron. It did not warp when heated, but slightly expanded, and was not porous (*Chem. News*, Vol. XVIII., p. 133, and XXI., p. 137.; also *Telegraphic Journal*, Vol. II., p. 128).

All ordinary depositing solutions of ferrous salts should be protected as much as possible from contact with the atmosphere, because they oxidise ; and a portion of the current subsequently passed through them is expended in deoxidising them. The oxidation is retarded by admixture of glycerine, which diminishes their diffusive power. When iron is deposited from some of these solutions, and has attained a certain thickness, brilliant scales of the metal become detached and fall to the bottom of the liquid.

Electrolysis of Ferrate of Potassium.—I have deposited iron from an aqueous solution of this salt, formed either by igniting peroxide of iron very strongly for some minutes with caustic potash and saltpetre, and dissolving the product in water, or by making a very strong solution of caustic potash, immersing in it a large iron or steel anode, and a small copper or platinum cathode, and passing a strong current from 15 or 20 Smee cells through it until it acquires a deep amethystine or purple colour. By that time the cathode had obtained a coating of iron, which was in the state of a dark powder if the powder was too great, but had the appearance of white cast iron (or intermediate between that and the appearance of reguline deposited zinc) when the powder was sufficiently weak. The solution rapidly decomposes, becomes colourless, and deposits all its metal in the state of peroxide at the bottom of the vessel.

Electrolysis of Ferrocyanide of Iron.—M. R. Bœttinger dissolves 10 parts of ferrocyanide of potassium and 20 of sodio-potassic tartrate in 200 of water ; then adds a solution of three parts of ferric sulphate previously dissolved in 50 parts of water, and then, with constant stirring, adds drop by drop a solution of caustic soda, until the precipitate of Prussian blue is just all redissolved. The resulting solution may be used for depositing iron upon copper (*Chem. News*, Vol. XXXVI., p. 11).

I have observed that an anode of iron greatly resists the passage of a current into a solution of perfectly pure potassic cyanide ; and that if a current of sufficient electromotive force is employed gas is freely evolved from the iron, and a minute portion of the metal is dissolved. I have also noticed that if a very thin wire of silver or gold and one of bright iron be weighed, then the two twisted together and immersed in a solution of potassic cyanide contained in a closed bottle, and set aside for several months, the silver or gold wire has partly or entirely dissolved, whilst the iron has lost not any or scarcely any of its weight.

(For the electrolytic analysis of compounds of iron, see *Chem. News*, Vol. XXXVIII., p. 26 ; Vol. XLII., p. 331 ; Vol. XLVI., p. 105 ; and *Jour. Chem. Soc.*, 1877, Part I., p. 341 ; Vol. XXXVIII., 1880, p. 284 ; Vol. XL., 1881, p. 1,081 ; Vol. XLII., 1882, pp. 896 and 1,320.)

Separation of Manganese.—Mn. Atomic weight = 55·0. A cation. By the simple immersion of sodium amalgam in an acidulated solution of a salt of manganese, metallic manganese is deposited and alloys with the mercury (Roussin, *Chem. News*,

Vol. XIV., p. 27; Watts's "Dictionary of Chemistry, Vol. VI., p. 802). According to Phipson, magnesium deposits manganese upon itself by simple immersion in a neutral solution of a manganous salt (*Proc.* Royal Society, 1864, Vol. XIII., p. 216; *Chem. News*, Vol. IX., p. 219).

Manganese is not deposited in the metallic state by a separate current from its neutral or acid solutions, but as hydrated manganese peroxide. In very dilute solutions of this metal containing much nitric, or a mixture of nitric and sulphuric acids, permanganic acid is formed, and colours the liquid (C. Luckow, *Jour. Chem. Soc.*, Vol. XXXVIII., p. 284).

Formation of Peroxide of Manganese.—The electrolysis, by a separate current, of a solution of nitrate or acetate of manganese yields a peroxide at the anode (W. Wernicke, Watts's "Dictionary of Chemistry," Vol. VII., p. 899; *Jour. Chem. Soc.*, Vol. IX., p. 307). Solutions of salts of manganese yield peroxide at the anode; one composed of one part of manganese chloride dissolved in eight of water yields, with a platinum wire cathode, very beautiful alternate rings of purple green, golden yellow, and blue, surrounded by a broad belt of golden yellow. With a solution composed of one part of acetate of manganese and fifteen of water, one uniform tint is invariably produced, first golden yellow, then purple, then green (B. Böttger, *Pogg. Ann.*, Vol. L., p. 45).

Electrolysis of Manganous Fluoride.—$Mn.F_2$. Molecular weight $= 93·0$. I melted some fluoride of manganese in a platinum crucible, and employed two spirals of platinum wire as electrodes, and a current from six large Smee cells. The conduction was moderate, and gas was evolved from the anode. In a few minutes both the cathode and the crucible became quite rotten by the union of the deposited manganese with the platinum. The anode was not corroded. I also melted the same salt in a crucible of copper, and passed the current by means of a sheet platinum anode and sheet copper cathode during half an hour. The conduction was free, abundance of gas was evolved from the anode, but none from the cathode, and it ceased on stopping the current. The deposit on the cathode was black, and did not evolve hydrogen with dilute hydrochloric acid, and was therefore not metallic manganese. The crucible was much corroded at the line of surface of the liquid.

I also electrolysed a dilute solution of fluoride of manganese by a current from six Grove cells and electrodes of platinum. Much heat was evolved, gas was set free at the anode, and a film of black deposit formed upon the cathode. By similar treatment of a saturated solution of the salt, not containing any free hydrofluoric acid, a film of purple colour was instantly formed upon the anode, but it dissolved quickly, and did not colour the liquid. Gas came from both electrodes freely; the liquid also became heated. No solid deposit was obtained.

Electrolysis of Manganous Chloride.—$Mn.Cl_2$. Molecular weight = 126. By the simple immersion of an amalgam of sodium in a saturated aqueous solution of this salt, Giles deposited manganese upon the surface of the amalgam (*Phil. Mag.*, 4th Series, Vol. XXIV., p. 328). It produces a viscid amalgam of manganese and mercury. According to S. Kern, magnesium deposits only manganous oxide (*Jour. Chem. Soc.*, 1876, Part 2, p. 479).

Bunsen filled a porous cell with a hot, saturated, aqueous solution of this salt, placed it in a charcoal crucible containing hydrochloric acid to the same level, kept the whole arrangement hot, and passed a current from four Bunsen cells from the crucible to a platinum wire immersed in the manganous solution. Metallic manganese was easily and freely deposited; but if the density of the current at the cathode was reduced by any means, or the concentration of the solution diminished, black manganous manganic oxide alone was obtained (*The Chemist*, No. XL, August, 1854, p. 685; Watts's "Dictionary of Chemistry," Vol. II., p. 438).

Electrolysis of Manganous Sulphate.—$Mn.SO_4$. Magnesium deposited hydrated manganous oxide from a neutral solution of this salt, but from the same solution acidified it deposits metallic manganese (Commaille, *Chem. News*, Vol. XIV., p. 188).

(For the electrolytic analysis of compounds of manganese, see *Jour. Chem. Soc.*, 1877, Part 2, p. 924, Vol. XXXVIII., p. 284, Vol. XLII., 1882, pp. 896, 1320; *Chem. News*, Vol. XLVI., p. 105.)

Deposition of Chromium.—Cr. Atomic weight = 52·5. A cation. A slightly acidified solution of chromic chloride or other chromic salt yields with sodium amalgam an easily decomposable liquid alloy, which, when heated in a stream of hydrogen or vapour of naphtha, loses its mercury and leaves metallic chromium in a spongy state. The liquid turns green previous to reduction (Bunge, Watts's "Dictionary of Chemistry," Vol. VI., p. 816; Roussin, *ibid.*, Vol. VI., p. 449; Vincent, *Phil. Mag.*, 4th Series, Vol. XXIV., p. 328). Magnesium precipitates only the hydrated sesquioxide of chromium from a solution of chromous and chromic chloride (Commaille, *Chem. News*, Vol. XIV., p. 188).

By means of a current from six Grove cells with platinum electrodes, I electrolysed a strong solution of fluoride of chromium containing some free hydrofluoric acid and a little hydrochloric acid. The liquid soon became hot; no gas was liberated at the cathode, but chlorine and ozone were set free at the anode, which was not corroded. I also passed a current from five Smee elements, by means of electrodes of platinum, through some acid potassic chromate in a state of fusion. A deposit slowly formed upon the cathode.

By operating in a similar manner upon a concentrated solution of chloride of chromium as upon one of manganese, Bunsen deposited chromium readily. The deposit appeared like iron, but was less affected by damp air. It resisted the action of boiling nitric acid, but was soluble in hydrochloric or dilute sulphuric acid. It was friable, and presented a polished surface on the side next the cathode. On diminishing the density of the current a black powder was deposited, containing more oxygen in proportion as the current was decreased. Adding protochloride of chromium had the opposite effect; *i.e.*, it caused metallic chromium to be deposited (*The Chemist*, No. 11, Aug. 1854, p. 686; Watts's "Dictionary of Chemistry," Vol. II., p. 438).

(For the electrolytic analysis of compounds of chromium see *Jour. Chem. Soc.*, Vol. XLII., p. 896.)

Deposition of Uranium.—Ur. Atomic weight = 120 A cation. Magnesium deposits gold-coloured hydrated sesquioxide of this metal by simple immersion in an aqueous solution of the oxalate of uranium (Commaille, *Chem. News*, Vol. XIV., p. 188). Magnesium decomposes an aqueous solution of uranic nitrate with evolution of hydrogen, and produces uranic oxide (S. Kern, *Jour. Chem. Soc.*, 1876, Part 2, p. 479). I melted some fluoride of uranium in a platinum crucible, and added to the liquid some crystals of silicon; the salt was not decomposed.

Solutions of uranium in mineral acids are not precipitated by a current from two to four Meidinger-Pincus elements, but the nascent hydrogen reduces the uranic to uranous oxide. From neutral solutions it is separated in very small quantities of a yellow colour. Alkaline solutions containing acetic, citric, or tartaric acid or sugar also deposit by electrolysis small quantities of uranium. The deposited uranium does not readily dissolve in dilute acids (Schicht, *Chem. News*, Vol. XLI., p. 280).

I passed a current from six Grove cells, by means of platinum electrodes, through an aqueous solution of fluoride of uranium. Much gas, having the odour of ozone, was evolved at the anode, and the liquid became hot. On adding some aqueous hydrofluoric acid the conduction became very free, and more gas was liberated from both electrodes, but no solid deposit was formed.

I also fused some fluoride of uranium in a copper crucible, and passed a current from six Smee cells through it by means of a platinum wire anode, using the crucible as a cathode; a little gas was set free at the anode, and the crucible melted. A second trial was made, using a platinum crucible, and two

H

spirals of platinum wire as electrodes, and the current continued during one hour. Conduction was very free, much gas was evolved from the anode, but none from the cathode; a bulky deposit quickly formed upon the negative spiral, especially on the side towards the anode. The deposit weighed 43·66 grains, and consisted of hard jet black crystals. The anode was not corroded. In a third trial four Grove cells were employed, and a special apparatus devised and employed to collect the evolved gas; about five cubic inches were obtained. The crystals were not metallic uranium; they were insoluble in boiling water, but soluble in cold dilute hydrofluoric acid, without evolving gas. About one-fourth of the deposit consisted of a fine crystalline powder, nearly of the colour of copper, but darker, and was composed of the crystals, with a film of less reduced fluoride upon them; they evolved gas in cold nitric acid, or in hot dilute nitric acid. They were not fused by heating alone to redness upon platinum foil, but if causticypotash was added they oxidised. I also electrolysed a fused mixture of the pure fluorides of uranium and potassium with platinum electrodes. The results were very similar, except that the deposit upon the cathode fell off as fast as it was formed, and the crystals had to be extracted by dissolving the cooled saline mass in slightly diluted and hot hydrochloric acid. They were very much like those of silicon; their form was that of a short pyramid with a square base. The anode was very slightly corroded, and made bright by the action, and twenty cubic inches of gas were collected from it.

By electrolysis with a separate current uranium is obtained in small quantity only, even from the completely neutral solution of the oxide, as a yellowish grey metallic precipitate, soluble in hydrochloric acid (C. Luckow, *Jour. Chem. Soc.*, Vol. XXXVIII., 1880, p. 284).

On passing the current from two elements of a bichromate of potassium battery through an aqueous solution of uranium acetate, formiate or nitrate, bright yellow uranium oxide, Ur_8O_4, was separated at the cathode, and gradually became black. No uranium remained in the solution after the current had been passed two hours. The black compound was uranic uranous oxide, containing 81·13 per cent. of uranium (E. F. Smith, *Chem. News*, Vol. XLIII., p. 61, also *Jour. Chem. Soc.*, Vol. XXXVIII., p. 284, Vol. XL., 1881, p. 3). According to the same author, molybdenum, tungsten, vanadium, didymium, and cerium are not completely precipitated from their solutions by the voltaic current.

(For the electrolytic analysis of compounds of uranium see *Chem. News*, Vol. XLII., p. 331.)

Separation of Tungsten. —W. Atomic weight = 184. A cation. When tungsten trioxide solutions are reduced by zinc, the final product of the action is tungsten dioxide (O. Freih, *Jour. Chem. Soc.*, 1883, Vol. XLIV., p. 785). I fused some sodic tungstate to a clear liquid in a porcelain vessel, and electrolysed it by means of a current from five Smee elements, a gas carbon anode, and a platinum wire cathode. The conduction was moderately free, gas was evolved from the anode, and at the cathode black matter was set free, floated, diffused in the liquid, and became partly redissolved. According to E. F. Smith, neutral solutions of the tungstates are not affected by the current (*Chem. News*, Vol. XLIII., p. 61).

Separation of Vanadium.—Va. Atomic weight = 137. L. Schicht dissolved vanadium chloride in water containing hydrochloric acid, and electrolysed the solution. No deposition took place in the blue liquid, the vanadic acid being merely reduced to oxide (*Chem. News*, Vol. XLI., . 280, and XLII., p. 331). I electrolysed a solution composed of vanadic acid dissolved in pure dilute hydrofluoric acid, by means of a current from 10 Smee elements, with a gas carbon anode and platinum cathode. Gas, having an odour of ozone, was set free at the anode. I also saturated dilute sulphuric acid with pure vanadate of ammonia, and electrolysed the solution with platinum electrodes, and a current from four zinc and platinum elements excited by dilute sulphuric acid. Conduction was very sparing; the solution slowly became of a very intense bluish black colour at the cathode, and a jet black powder of some thickness was deposited upon it.

(For the electrolytic analysis of vanadium compounds see *Jour. Chem. Soc.*, Vol. XLII., 1882, p. 896.)

Separation of Molybdenum. — Mo. Atomic weight = 96. A cation. Sodium molybdate is not reduced by metallic tin (Ullik, Watts's "Dictionary of Chemistry," Vol. VI., p. 832). From an ammoniacal solution of molybdic anhydride, by means of a separate current, molybdenum is completely and firmly deposited upon the cathode as molybdous oxide in coloured rings which thicken and become black. The first blue precipitate is molybdic molybdate, then follow molybdic and molybdous oxides. In acid solutions there is no deposit. In ammonium molybdate acidified with molybdic anhydride the precipitation is incomplete (L. Schicht, *Jour. Chem. Soc.*, Vol. XXXVIII., 1880, p. 747; *Chem. News*, Vol. XLI., p. 280, and Vol. XLII., p. 331).

Molybdic acid dissolves freely in pure dilute hydrofluoric acid, evolving a little heat. I electrolysed the solution both with a platinum and with a gas carbon anode and a current from ten large Smee cells. The colourless liquid conducted freely, becoming instantly of an indigo blue colour at a platinum cathode. Gas was set free at each electrode; that from the carbon anode was the most abundant, and had a slightly chlorous odour. On stopping the current the deep blue film on the cathode quickly dissolved, and the liquid soon became colourless. During the action the cathode was several times removed and dipped into water; much blue matter dissolved, but the water became nearly colourless in half a minute, even without stirring, and however large the quantity of the blue matter was which dissolved it.

I also fused some molybdic acid in a porcelain crucible, and passed a current through it from five Smee elements, by means of a gas carbon anode and platinum cathode. It conducted freely. The action was rather strong at the anode, but little gas was set free. No gas was evolved at the cathode, but crystals quickly collected upon it in a large mass, which soon filled the entire solution and spread to the anode. The carbon was not disintegrated or dissolved. The cooled residue was a black mass of crystals. In a second trial with a current from 12 similar cells, and a platinum anode and cathode, much gas was set free at the cathode, and less from the anode, and the bluish black deposit formed upon the cathode. A large number of crystalline needles, from $\frac{1}{8}$th to $\frac{1}{4}$th of an inch long, stood out at right angles upon the surface of the cathode in the liquid. The deposit imparted a transient green colour to water.

Crystals of dioxide of molybdenum, MoO_2, quickly become covered with copper when immersed in a solution of cupric sulphate in contact with zinc (Ullik, Watts's "Dictionary of Chemistry," Vol. VI., p. 832).

(For the electrolytic analysis of molybdenum compounds *see* E. F. Smith, *Chem. News*, Vol. XLIII., p. 6; and also *Jour. Chem. Soc.*, Vol. XL., 1881, p. 3.)

Separation of Lead.—Pb. Electro-chemical equivalent $\frac{207}{2} = 103\cdot5$. A dyad cation. The deposition of lead by the simple immersion of zinc in a solution of nitrate or acetate of lead is a very old fact, and when the zinc is in the form of a spiral wire it constitutes the well-known "lead tree." According to A. Cossa an alkaline solution of plumbic chromate is at once decomposed by aluminium, with deposition of lead and formation of chromic oxide (Watts's "Dictionary of Chemistry," Vol. VII., p. 54). Thallium deposits lead from a solution of plumbic acetate (W. C. Reid, *Chem. News*, Vol. XII., p. 242). Lead in contact with gold in acid or neutral, cold or hot, solutions of salts of lead, produces no deposit of lead (Raoult, *Jour. Chem. Soc.*, Vol. XL, p. 646).

Electrolysis of Plumbic Nitrate.—Pb2NO₃. Molecular weight = 331. A solution of this salt is slowly decomposed by contact with aluminium, and the lead deposited in crystals (A. Cossa, Watts's "Dictionary of Chemistry," Vol. VII., p. 54). Magnesium immersed in a solution of plumbic nitrate

is quickly covered with lead powder, which quickly oxidises.
(S. Kern, *Jour. Chem. Soc.*, 1876, Part I., p. 683). In a solution of hyponitrite, nitrate, or acetate of lead, zinc received a
coating of lead by simple immersion, but antimony, bismuth,
tin, lead, iron, nickel, copper, brass, German silver, silver,
gold, or platinum did not. A solution of the nitrate yields by
electrolysis with a separate current peroxide of lead at the
anode.

Electrolysis of Plumbic Fluoride.—PbF_2. I fused some
of this salt in a platinum crucible, and added some crystals of
boron ; vivid incandescence occurred, and melted lead was
separated. The addition of crystals of silicon had a similar
effect. Metallic antimony or copper did not liberate lead. By
stirring the melted salt with an iron rod heat was evolved, the
iron corroded, and lead was set free. The addition of zinc to
the fused salt caused an explosion, and magnesium produced
quite a dangerous detonation.

Beetz electrolysed this salt in a fused state by a separate
current, and observed that a colourless gas was evolved from
the anode, and lead set free at the cathode (*The Chemist*,
New Series, Vol. I., p. 253). G. J. Knox electrolysed it with
an anode of charcoal and a cathode of platinum wire by means
of a current from sixty voltaic cells (*Phil. Mag.*, 3rd Series,
Vol. XVI., p. 192). Fremy found it easily decomposed by a
separate current; lead was deposited (*The Chemist*, New Series,
Vol. II., p. 548). I melted 400 grains of the pure salt in a
thick copper crucible, and electrolysed the liquid by means of
a current from six Smee cells, using a platinum wire anode
and a copper wire cathode. Conduction was copious; a bulky
crust quickly formed upon the cathode, and advanced towards
the anode in lumpy projections. A little gas appeared at the
latter, but during a short time only. The deposit upon the
cathode was not lead, nor was there any metal contained in a
free state in it, nor in the saline mass, after action lasting one
hour ; it was a mass of lead salt, brittle and of a red-brown
colour (like that of peroxide of lead) when cold. The conduction was very perfect, and the fused salt appeared to conduct
without being decomposed. The anode was not corroded. I
also electrolysed the fused salt in a deep, narrow, and thick
copper cup, with an anode of gas carbon, during one and
a-quarter hour. The anode was corroded, and the metal
liberated ; action was copious, gas was evolved at the anode,
and about seven or eight cubic inches were collected.

Electrolysis of Plumbic Chloride.—$Pb.Cl_2$. Molecular-
weight $= 278$. According to Commaille, magnesium deposits
lead, together with much hydrogen, from a neutral solution of
this salt (*Chem. News*, Vol. XIV., p. 188). Aluminium immediately deposits crystals of lead from it (A. Cossa, Watts's

"Dictionary of Chemistry," Vol. VII., p. 54). According to Becquerel, if a piece of bright copper in contact with zinc be immersed in a solution of the chlorides of lead and sodium it becomes coated with lead (*The Chemist*, Vol. V., p. 408).

Faraday found that the proportion of lead deposited from its fused chloride to that of water decomposed by the same current was as 100·85 to 18 (Watts's "Dictionary of Chemistry," Vol. II., p. 439). According to Buff, solid lead chloride conducts like a metal—*i.e.*, without decomposition—but rise of temperature *increases* its conductivity (*Jour. Chem. Soc.*, 1876, Part I., p. 668). Faraday found that by passing a current through the melted salt chlorine appeared at the anode and lead at the cathode.

Electrolysis of Plumbate of Potash.—Metallic zinc or tin, but not iron, becomes coated with lead by simple immersion in a solution formed by dissolving litharge in a boiling hot solution of caustic potash.

Haeffelly deposits lead upon copper or brass by immersing them in contact with a piece of tin in a hot alkaline solution of oxide of lead. The tin dissolves in the form of an alkaline stannate, and the lead is deposited in a spongy state (*Chem. News*, Vol. VI., p. 163). I connected together a wire of zinc and one of platinum, and immersed the pair in a solution of litharge in strong aqueous ammonia; both wires became coated with a black deposit in a few minutes. By contact with air, the moist deposit became yellow, and was apparently reconverted into litharge. F. Weil coats copper, iron, and steel with lead, by dissolving a salt of lead in a strong solution of potash or soda, and immersing them in the liquid in contact with zinc; the deposit, however, contains zinc. To obtain it pure, the piece of zinc is placed in the alkaline lixivium in a porous cell, and the cell immersed in the lead solution, the zinc being connected with the copper, &c., by a wire (*Chem. News*, Vol. XIII., p. 2).

Electrolysis of Plumbic Acetate.—According to A. Cossa, aluminium slowly deposits lead in crystals from a solution of this salt (Watts's "Dictionary of Chemistry," Vol. VII., p. 54). By a separate current, this solution yields peroxide of lead at the anode.

Formation of Peroxide of Lead.—According to W. Wernicke, an alkaline solution of the tartrate of lead and sodium, with platinum electrode and a current from two Daniell cells, yields a black deposit of peroxide of lead upon the anode; and a solution of one part of plumbic nitrate and eight of water gives a similar deposit by such treatment (*Jour. Chem. Soc.*, Vol. IX., p. 306 ; *Chem. News*, Vol. XXII., p. 240).

Nobili, in the year 1826, discovered that if a solution of acetate of lead be electrolysed by means of a large sheet platinum anode and a platinum wire cathode, a deposit is formed upon the positive plate ; and that if a polished steel plate be employed as the anode, with a current from four or six Grove cells, the deposit is in the form of a thin film, and exhibits all the colours of the spectrum ; and by placing the positive plate horizontally beneath the vertical negative wire the colours are in the form of rings, the centre of which is the wire, and are arranged in the order of the chromatic scale. These colours are known as " Nobili's rings." Becquerel, Gassiot and others have, by varying the strength of the battery and of the solutions employed, and interposing non-conducting patterns between the anode and cathode, and by using cathodes of different shapes, obtained effects of great delicacy and beauty. Salts of other metals, such as bismuth, silver, nickel, cobalt, manganese, &c., which yield deposits of peroxide at the anodes, may be employed instead of those of lead. Becquerel prepared his plumbic solution as follows :— Dissolve 200 grammes of caustic potash in two quarts of distilled water, add 150 grammes of litharge, boil the mixture half an hour, allow it to become clear, take the clear portion and dilute it with its own bulk of water (*The Chemist*, Vol. IV., p. 457). The solution is used cold, and is rapidly deprived of its metal, because lead is deposited upon the cathode at the same time.

By this means may be imparted to polished surfaces of metals all the richest colours of the rainbow. "They commence with silver blonde, and progress onwards to fawn colour, and thence through various shades of violet to the indigo and blues; then through pale blue to yellow and orange; thence through lake and bluish lake to green and greenish orange and rose orange; thence through greenish violet and green to reddish yellow and rose lake, which is the highest colour on the chromatic scale " (Walker's " Electrotype Manipulation," Part XL, 16th edition, p. 40). Too great a strength of the current covers all the tints with an uniformly dark brown coating. The deposits, if properly prepared, resist friction well. The process is termed " Metallo-chromy."

Metallo-chromy effected by means of a solution of oxide of lead in caustic soda, or potash, is largely employed in Nuremburg to ornament metallic toys (Wagner's " Technology," p. 117). Bells are similarly coloured in France, and the hands and dials of watches in Switzerland.

Electrolytic Analysis of Compounds of Lead.—See *Jour. Chem. Soc.*, Vol. XXXVIII., 1880, p. 284 ; Vol. XLII., 1882, p. 1,320 ; *Chem. News*, Vol. XXXV., p. 264; Vol. XLVI., p. 106; Watts's " Dictionary of Chemistry," Vol. VIII., Part I., p. 712, Part II., p. 1,168.

(For Keith's process for desilvering lead by means of an electric current, see *Jour. Chem. Soc.*, 1877, Part XI., pp. 804

and 924; Vol. XXXVI., 1879, pp. 288 and 410; and for Blagden's process, *see* Watts's "Dictionary of Chemistry," Vol. VI, p. 1,026)

Separation of Thallium.—Tl. Electro-chemical equivalent = 204. A monad cation. Zinc coats itself with metal in solutions of salts of thallium, but tin usually does not. According to Lamy, zinc precipitates the metal from the solutions of the nitrate and sulphate in the form of brilliant crystalline laminæ. I found that crystals of silicon had no reducing effect on a solution of fluoride of thallium containing free hydrofluoric acid. According to A. Cossa, aluminium deposits by simple immersion metallic thallium from a solution of thallium chloride at 90° C. (Watts's "Dictionary of Chemistry," Vol. VII., p. 54).

Solutions of the salts of this metal are easily decomposed by a feeble current, and the metal deposited in beautiful crystalline plates upon the cathode. I electrolysed an aqueous solution of the fluoride by a current from a single Smee element, a thallium anode and a platinum cathode. It conducted freely, and quickly gave a metallic deposit, in long feathery crystals, like those of electro-deposited tin, but of a less white colour.

According to L. Schicht, acidulated solutions of nitrate and sulphate of thallium were not precipitated by a separate current. From ammoniacal solutions thallium was deposited upon the cathode together with much gas; whilst at the anode there appeared blackish brown thallium oxide much resembling peroxide of lead. The current was from four Meidinger-Pincus elements, and yielded the metal in a spongy state and of a dark colour; but by using only two or three cells, fine permanently adhesive metal was obtained. From neutral solutions the metal is imperfectly precipitated on account of the acid which is liberated, but in alkaline ones the metal is bright and solid, and the deposition is complete. The deposit redissolves readily in sulphuric acid (*Chem. News*, Vol. XLI., p. 280; also Vol. XLVII., p. 209).

Electrolysis of Sulphate of Thallium.—Tl_2SO_4. Molecular weight = 504. Aluminium immersed in a slightly acid solution of thallium sulphate becomes coated in ten days with regular octohedra of thallium alum (Watts's "Dictionary of Chemistry," Vol. VII., p. 54). A solution of sulphate of thallium, acidulated with sulphuric acid, deposits its metal upon zinc by simple immersion (*Chem. News*, Vol. XXXVI., p. 166).

A thallium anode, in water acidulated with sulphuric acid, is converted into the black trioxide by a current from two Bunsen cells (Watts's "Dictionary of Chemistry," Vol. VI., p. 1,082).

The metal is reduced from its solutions, generally from the sulphate, either by a separate current or by simple immersion of zinc. When a current from two or three Grove cells, with platinum electrodes, is passed through an acidulated solution of thallium sulphate, dissolved in its own weight of water, the metal is deposited upon the cathode in brilliant plates and long needle-shaped crystals stretching out towards the anode. The reduction is complete when hydrogen begins to escape at the cathode (Crookes, Watts's " Dictionary of Chemistry," Vol. V., p. 743).

(For the electrolytic analysis of compounds of thallium, *see* L. Schicht, *Chem. News*, Vol. XLII., p. 331.)

Separation of Indium.—In. Atomic weight = 113·4. A cation. This metal being very costly, but little has been done with it in electro-chemistry. It is deposited from solutions by simple immersion of zinc in them.

According to L. Schicht, indium is completely deposited as a bluish white metal at the cathode by a separate current, both from acid and from alkaline solutions. In the latter case the metal is very bright and firm. With solutions containing organic acids, indium is also deposited in a coherent state, with abundant escape of gas (*Chem. News*, Vol. XLI., p. 280 ; also Vol. XLVII., p. 209).

(For the electrolytic analysis of compounds of indium, *see* L. Schicht, *Chem. News*, Vol. XLII., p. 331.)

Separation of Tin.—Sn. Atomic weight = 118. A cation. Very few solutions of tin are available for electrolysis; the chief are stannous chloride and the aqueous solutions of stannous and stannic fluoride.

M. H. Loewel added metallic tin to a solution of green crystallised chloride of chromium free from excess of acid, in a closed glass vessel, and boiled the mixture during about 10 or 12 minutes, and then allowed it to cool. During the heating the tin combined with the chlorine of some of the chromium salt, forming stannous chloride and chromous chloride ; but during the cooling the action was reversed, the chromous chloride took chlorine from the stannous chloride, and metallic tin was deposited in the form of numerous small plates (*The Chemist*, Part VIII., May, 1854, p. 476).

Electrolysis of Stannous Fluoride.—$Sn.F_2$. Molecular weight = 156. I found that zinc, immersed in a solution of stannous fluoride, produced a flocculent precipitate, and evolved gas. In the same liquid, but containing free hydro-fluoric acid, crystals of silicon did not deposit tin by simple immersion. Fremy electrolysed fused fluoride of tin in a platinum vessel ; it was easily decomposed ; the deposited metal alloyed with and perforated the vessel in a few minutes (*The Chemist*, New Series, Vol. II., p. 548). I electrolysed a saturated non-acid solution of stannous fluoride by means of large platinum electrodes, and a current from 10 large Smee cells; the conduction was sparing, a little oxygen was evolved at the anode, and long feathery crystals of tin were slowly formed upon the cathode. No gas appeared at the cathode or solid deposit at the anode. By using only one Smee cell the deposit of tin was white, and beautiful crystals of the metal soon reached across the liquid, and completed the metallic circuit by touching the anode.

Also by passing a current from six Grove elements by means of platinum electrodes through a strong aqueous solution of *stannic* fluoride containing little or no free hydrofluoric acid a grey deposit of tin soon appeared on the cathode. The conduction was free, much gas came from the anode, and heat was evolved in the liquid. The anode was not corroded, nor did it acquire any solid deposit.

Electrolysis of Stannous Chloride.—Sn.Cl$_2$. Molecular weight = 189. Electrolytic experiments for the separation of tin are usually made with solutions containing this salt. Magnesium deposits stannic acid and spongy tin from this solution (Commaille, *Chem. News*, Vol. XIV., p. 188). A "tin tree" is produced by immersing a spiral of zinc wire in ten to twenty ounces of water in which has been dissolved three drachms of this salt and ten drops of nitric acid, and allowing the arrangement to remain undisturbed. According to Böttger, sodium amalgam in contact with a concentrated solution of stannous chloride forms a viscid amalgam. Joule obtained a beautiful crystalline amalgam by using a separate current and making mercury the cathode in this liquid. I have observed that zinc and lead become tinned by simple immersion in a solution of the salt, but antimony, bismuth, platinum, gold, silver, copper, brass, German silver, nickel, iron, and tin do not. According to Raoult, gold or copper in contact with tin in a concentrated and boiling solution of stannous chloride receive a deposit of tin ; but gold in contact with antimony, silver, copper, nickel, iron, or lead receives no such coating in either the hot or cold liquid (*Chem. News*, Vol. XXVI., p. 240, and XXVII., p. 59 ; also *Jour. Chem. Soc.*, Vol. X., p. 464).

Zinc or iron previously coated with a film of metallic copper by simple immersion process acquire a deposit of tin by simple contact with a solution composed of one part of crystals of stannous chloride, two of water, and two of hydrochloric acid (C. Paul, *Jour. Chem. Soc.*, Vol. XL, p. 955). According to Roseleur, zinc and iron become tinned by simple immersion in a boiling hot solution composed of one part of fused stannous chloride, thirty of ammonium alum, and 2,000·of water; zinc also acquires a coating of tin by simple contact with a solution of one part of fused stannous chloride and five of pyrophosphate of sodium, dissolved in 300 parts of distilled water.

Copper, brass, and bronze become coated with tin by contact during a few minutes with that metal in a boiling hot solution of peroxide of tin in caustic potash. F. Weil coats copper with tin by immersing it, in contact with zinc, in a solution formed by dissolving a salt of tin in a strong solution of caustic potash or soda, the liquid being at 50° to 100°C.; the deposit, however, contains zinc (*Chem. News*, Vol. XIII., p. 2). Dr. Hillier tins metals by immersing them in contact both with tin and zinc in a hot solution of one part of stannous chloride dissolved in 20 of water, to which has next been added one or two parts of caustic soda in 20 of water.

According to Becquerel, copper and iron become tinned by immersion in contact with zinc in a dilute solution of the double chloride of tin and sodium at 160° F., but are not tinned by simple immersion alone in that liquid (*The Chemist,* Vol. V., p. 408). For coating iron with tin by immersion in a liquid in contact with zinc Roseleur recommends a solution prepared thus:—Take equal weights of stannous chloride, cream of tartar, and water. Dissolve the chloride in one-third of the cold water, warm the other portion of water and dissolve the cream of tartar in it, and mix the solutions ; the mixture is clear, and has an acid reaction. And a second solution, composed of six parts of crystal, or four of fused stannous chloride, and 60 of pyrophosphate of potassium or sodium, dissolved in 2,000 parts of distilled water. The size of the zinc should be about $\frac{1}{15}$ that of the iron. The deposition occupies several hours. When the solution becomes weak equal weights of the pyrophosphate and fused chloride are added.

M. Heeren coats iron with tin by immersing it during two hours, in contact with zinc, in a solution of two parts of tartaric acid, three of stannous chloride, and three of caustic soda, and 100 of water (*Jour. Chem. Soc.*, Vol. XIII., p. 672). Stolba uses a solution of 5 to 10 parts of stannous chloride dissolved in 100 of water, and a very minute amount of cream of tartar added. The metal to be coated is wetted with the solution whilst in contact with particles of zinc spread over its surface (*Chem. News*, Vol. XXIII., p. 21). Brass and copper acquire a coating of tin if placed in contact with that metal in a boiling hot saturated solution of cream of tartar.

By means of the single cell process F. Weil coats copper with tin in a solution of a salt of tin in strong caustic potash or soda. A porous cell, containing a solution of the potash or soda, is placed in the bath, a piece of zinc immersed in it, the copper immersed in the hot tinning liquid, and the two metals connected together by a wire. The deposit is pure tin, and may be obtained of any thickness. To revive the inner liquid, precipitate the dissolved zinc from it by addition of solution of sulphide of sodium (*Chem. News*, Vol. XIII., p. 2).

By means of a separate current, fused stannous chloride yields tin at the cathode, whilst vapour of stannic chloride escapes at the anode (Faraday). He found by experiment that the proportion by weight of tin deposited from fused stannous chloride, and of water decomposed by the same current was as 117·16 to 18 (Watts's "Dictionary of Chemistry," Vol. II., p. 439). Iron may be quoted with a beautiful white deposit of tin by making it the cathode in a solution of stannate of potash; but the solution is gradually decomposed by contact with the atmosphere, and deposits peroxide of tin.

Various solutions yield tin by this method. Roseleur's is composed of six parts of crystals of stannous chloride, and 50 of pyrophosphate of sodium, dissolved in 5,000 parts of distilled water, the two salts being dissolved in separate portions of the water, and the solutions mixed, and then stirred till clear. It requires a large anode and a strong current. (For various other electrolytic mixtures containing stannous chloride and other ingredients, *see* "The Art of Electro-Metallurgy," Longman's "Text-Books of Science," pp. 270–272.)

Anhydrous stannic chloride did not conduct a current from 8,040 cells of W. de la Rue's chloride of silver battery (Bleekrode, *Proc. Roy. Soc.*, Vol. XXV., p. 325).

Separation of Alloys of Copper and Tin.—Iron is said to acquire a deposit of bronze by simple immersion in a solution of 4 to 5 parts of cupric sulphate, 4 to 5 of crystallised stannous chloride, and 100 of water.

For depositing bronze by a separate current, Salzede used a solution composed of cupric chloride, stannous chloride, nitrate of ammonium, and potassic carbonate and cyanide, dissolved in water. For the same purpose Newton used one composed of the tartrates of copper, tin, and potassium.

Formation of Crystals of Tin by Electrolysis.—The crystallisation of tin is a phenomenon conspicuously striking under some conditions in a solution of stannous chloride. The crystals of tin formed upon the cathode increase so rapidly in length as to grow across the solution, and touch the positive pole in a few minutes. And if the solution and current are strong and the cathode small, quite a mass of crystals will soon fill the liquid and converge towards the anode. If the anode be drawn farther away in the solution the crystals follow it. The largest crystals are produced by slow action; to produce them a platinum capsule is covered with an outer coating of wax, leaving the bottom uncovered, and then set upon a plate of amalgamated zinc in a porcelain vessel. The capsule is then filled completely with a dilute and not too acid solution of stannous chloride, whilst the outer vessel is filled with water (containing one-twentieth its bulk of hydrochloric acid) up to such a height that the two liquids come into mutual contact. The electric current generated reduces the salt of tin, and in a few days the crystals upon the interior of the capsule are well developed, and should be washed with water and dried quickly (F. Stolba, *Chem. News*, Vol. XXX., p. 177).

For the electrolytic analysis of compounds of tin see *Chem. News*, Vol. XLVI., p. 106. Also Watts's "Dictionary of Chemistry," Vol. VI., p. 676; *Jour. Chem. Soc.*, Vol. XL., 1881, p. 1,081, Vol. XLII., 1882, p. 1,320.

Separation of Cadmium.—Cd. Electro-chemical equivalent $=\dfrac{112}{2}=56$. A dyad cation. Sodium amalgam decomposes a solution of a salt of cadmium, and forms cadmium amalgam (Böttger). From a solution of the chloride, magnesium deposits, with strong action, a mixture of cadmium and an oxy-chloride of the same metal (Commaille, *Chem. News*, Vol. XIV., p. 188). I have found that crystals of silicon heated with cadmic fluoride set free cadmium.

According to Raoult, gold or copper in contact with cadmium in a concentrated and boiling solution of cadmium sulphate or chloride decomposes these salts, and quickly deposits a white, brilliant, and firmly adherent but thin film of cadmium upon the gold or copper, even when the solution is not acidulated and no hydrogen evolved. The experiment does not succeed with the nitrate. But gold in contact with iron, nickel, antimony, lead, copper, or silver, in cold or boiling acid or neutral solutions of salts of cadmium, receives no such deposit (*Chem. News*, Vol. XXVI., p. 240; Vol. XXVII., p. 59; *Jour. Chem. Soc.*, Vol. XL, p. 464).

By means of a separate current a spongy deposit of cadmium is obtained from its chloride solution to which a few drops of sulphuric acid have been added. Cadmium ammonio chloride gives a grey non-adherent deposit, chlorine being evolved. A similar deposit was obtained from cadmium calcium chloride. Cadmium bromide, acidulated with weak sulphuric acid, gives a coherent mass, susceptible of polish. If an iron wire be used as the cathode, and a copper one as the anode, the cadmium is deposited in long brilliant needles. A good result is also obtained with cadmium ammonio bromide. Cadmium ammonium iodid, yields a spongy mass. The sulphate gives a coherent deposit capable of receiving a fine polish. A non-coherent deposit was obtained from the double sulphate of cadmium and ammonium (A. Bertrand, *Jour. Chem. Soc.*, 1887, Part I., p. 161). Russell and Woolrich deposited cadmium by the electrolysis of a solution composed of cadmic carbonate dissolved in aqueous potassic cyanide, with free cyanide added, and using the liquid at about 100° F. with a cadmium anode.

For the electrolytic analysis of compounds of cadmium, see Jour. Chem. Soc., 1877, Part L, p. 340. Also F. Beilstein, Vol. XXXVI., 1879, pp. 276 and 746; Vol. XL., 1881, p. 1,081; Vol. XLII., 1882, p. 8,960. Watts's "Dictionary of Chemistry," Vol. VII., pp. 229 and 790 (E. J. Smith). Chem. News, Vol. XXXIX., p. 185; ditto (Beilstein and Jamain), Vol. XL., p. 109; Vol. XLIII. (E. Smith) p, 61; and V. Francken, Vol. XLVI., p. 106. Jour. Chem. Soc., Vol. XLII., 1882, p. 1,320.

Separation of Zinc. — Zn. Electro-chemical equivalent $= \dfrac{65}{2} = 32\cdot5$. A dyad cation. Only the most highly positive metals usually set free zinc from its solutions. From slightly acid solutions of zinc salts magnesium deposits the metal and hydrogen gas (Roussin, Chem. News, Vol. XIV., p. 27). According to S. Kern, magnesium evolved hydrogen very slowly from a solution of zinc chloride (Jour. Chem. Soc., 1876, Part I, p. 684). Sodium amalgam immersed in a concentrated solution of zinc sulphate forms a viscid amalgam of zinc (Böttger, Watts's "Dictionary of Chemistry," Vol. III., p. 891). Joule also obtained amalgams of zinc by electrolysis, using a cathode of mercury (ibid.). From an alkaline solution of a salt of zinc aluminium easily separates the metal (A. Cossa, Watts's "Dictionary of Chemistry," Vol. VII., p. 54). I observed that in a solution of either nitrate, chloride, sulphate, or acetate of zinc neither antimony, bismuth, platinum, gold, silver, copper, brass, German silver, nickel, iron, tin, lead, or zinc becomes coated with zinc by simple immersion. I heated a mixture of 1·5 grain of crystals of silicon and 10·25 grains of perfectly dry fluoride of zinc in a porcelain crucible to a full red heat; the salt was decomposed and zinc set free. According to V. Roque, wrought and cast iron previously dipped in a strong solution of potassic carbonate became coated with zinc by simple immersion during from three to twelve hours in a solution composed of 1,000 parts of water, 10 of chloride of aluminium, eight of potassic bitartrate, five of stannous chloride, four of acid sulphate of aluminium, and four of chloride of zinc (Chem. News, Vol. XXI., p. 288).

Raoult states that gold or copper in contact with zinc, in a concentrated and boiling solution of chloride or sulphate (but not nitrate) of zinc, acquires a deposit of zinc. But gold in contact with antimony, silver, copper, nickel, iron, or lead, in cold or boiling acid or neutral solutions of salts of zinc, receives no such coating (Chem. News, Vol. XXVI., p. 240; Vol. XXVII., p. 59. Jour. Chem. Soc., Vol. XL, p. 464). Copper or brass immersed in contact with zinc in a boiling saturated solution of chloride of ammonium acquires in a few minutes a specular coating of zinc, but in a solution of cream of tartar no. such deposit occurs (R. Böttger, "Gmelin's Handbook of Chemistry," Vol. L, p. 50; also *Chem. News*, Vol. XXII., p. 108). Copper acquires a fixed and brilliant coating of zinc by immersing it in contact with zinc in a hot concentrated solution of potash or soda (F. Weil, *Chem. News*, Vol. XIII., p. 2).

By means of a separate current and a zinc anode zinc has been deposited from solutions of several of its salts, viz., the chloride, ammonio chloride, sulphate, ammonio sulphate, acetate, tartrate, &c. According to Smee, a solution of zinc oxide in caustic potash is not a good conductor; the zinc anode does not readily dissolve in it, and similarly with potassio tartrate and potassio of cyanide.

Electrolysis of Chloride of Zinc.—$Zn.Cl_2$. Molecular weight = 136. Fused zinc chloride is reduced to metal by contact with aluminium (Flavitzky, Watts's "Dictionary of Chemistry," Vol. VIII., Part L, p. 64). It has been stated that perfectly clean iron acquires a thin coating of zinc by simple immersion in a solution of 30 parts of zinc chloride and 1 of sal-ammoniac (Watts's "Dictionary of Chemistry," Vol. VIII., Part IL, p. 1,118). According to Grove, nitride of zinc is formed at an anode of zinc in a weak solution of sal-ammoniac (Watts's "Dictionary of Chemistry," Vol. V., p. 1,072).

Electrolysis of Sulphate of Zinc.—$Zn.SO_4$. Molecular weight = 161. Sodium amalgam in contact with a strong solution of this salt forms a viscid amalgam (Böttger). Joule formed the same compound by making mercury the cathode in that liquid. From a solution of the sulphate, magnesium deposits with strong action a mixture of zinc, its hydrated oxide, and sulphate (Commaille, *Chem. News*, Vol. XIV., p. 188).

By means of a current from two Smee cells, with a large zinc anode, a solution of one part of zinc sulphate in five to ten parts of water may be made to yield a good deposit of zinc. According to V. Meyer, pure zinc may be obtained by the electrolysis of an ammoniacal solution of its sulphate with a sheet zinc anode and a copper wire cathode (*Jour. Chem. Soc.*, Vol. X., 2nd Series, p. 221; *see* also Watts's "Dictionary of Chemistry," Vol. VII., p. 1,213).

MM. Person and Sire easily deposited zinc "on any metal," by the separate current process, with a single cell and a zinc anode, from a solution of one part of oxide of zinc dissolved in 100 parts of water containing 10 of alum (*Chem. News*, Vol. II., p. 275).

Electrolysis of Cyanide of Zinc and Potassium.—A. Watt makes a mixture composed of twenty gallons of distilled water, 200 ounces of cyanide of potassium, and eighty by measure of the strongest aqueous ammonia. He then fills several large porous cells with a solution composed of sixteen ounces of cyanide of potassium to each gallon of water, and partly immerses them in the other liquid. In the porous cells he places sheets of copper or iron to act as cathodes, and in the outer liquid clean pieces of zinc to act as anodes, and connects the battery in the usual way until about sixty ounces of zinc are dissolved, and then stops the current and removes the porous vessels. He next dissolves eighty ounces of carbonate of potassium in a part of the zinc solution, and returns it to the original portion, and stirs the mixture thoroughly. After the sediment formed has subsided he decants the clear liquid for use. Articles of iron may be coated in this liquid. Anodes of zinc are employed, and a little cyanide of potassium and liquid ammonia are occasionally added if necessary. The battery preferred is composed of two Bunsen cells.

Deposition of Alloys of Zinc and Copper.—As early as the year 1841 M. de Ruolz deposited brass, by means of the battery process, from a solution of the mixed cyanides of copper, zinc, and potassium. One of the best solutions for yielding brass by means of a separate current is that of Morris and Pershouse. It is composed of one pound of potassic cyanide, one of ammonium carbonate, two ounces of cupric cyanide, and one of cyanide of zinc, dissolved in one gallon of water, and the liquid used at 150° F., with a strong current and a large brass anode. To increase the proportion of copper in the deposit, either add potassic cyanide or raise the temperature, and to increase that of the zinc, add ammonic carbonate or lower the temperature. Walenn recommends a solution composed of equal parts of ammonic tartrate and potassic cyanide dissolved in water, and after addition of the cyanides of copper and of zinc the oxides of those metals are also added to the solution. If upon trial hydrogen is set free at the cathode, a little ammoniuret of copper is also added to the mixture. There is then no liberation of hydrogen, and a deposit of brass may be obtained of any desired thickness. Two or three Smee cells are sufficient (*Chem. News*, Vol. XXL, p. 273, Vol. XXII., pp. 1 and 181; *Jour. Chem. Soc.*, Vol. X., p. 103; *Phil. Mag.*, 4th Series, Vol. XLI., p. 41). In depositing from an electro-brassing solution, which contains cyanide of potassium and tartrate of ammonium, at a temperature but little above the freezing point, nearly pure zinc forms upon the cathode (Walenn, *Chem. News*, Vol. XXXV., p. 154; *see* also Watts's "Dictionary of Chemistry," Vol. VII., p. 382).

Deposition of Alloys of Zinc, Copper, and Nickel.—A solution described by Morris and Johnson for depositing German silver is composed as follows :—Dissolve one pound of potassic cyanide and one of carbonate of ammonium in a gallon of water. Heat the solution to 150° F. Immerse it in a large anode of German silver and a small cathode of any suitable metal, and pass a strong current until a large quantity of the alloy has dissolved and a bright cathode receives a good deposit of the desired alloy. If the deposit becomes too red, add ammonic carbonate ; if too much of the appearance of zinc, add potassic cyanide.

Various other mixtures for depositing brass and German silver may be found described in works on electro-metallurgy.

For the electrolytic analysis of compounds of zinc, see *Chem. News*, Vol. XXIV., pp. 100 and 172 ; Vol. XXXV., p. 264 ; Vol. XLIII., p. 61 ; Vol. XLIV., p. 304 ; and Vol. XLVI., p. 105. *Jour. Chem. Soc.*, 1877, Part L, p. 340, and Part II., pp. 804 and 924 ; Vol. XXXVIII., 1880, p. 584 ; Vol. XL., 1881, pp. 1,081, 1,101, and 1,170 ; Vol. XLII., 1883, pp. 896 and 1,320 ; and Vol. XLIV., 1883, p. 122. *The Chemist*, New Series, Part XVIII., March, 1855, p. 334. Watts's "Dictionary of Chemistry," Vol. VIII., Part I., p. 712.

For the electro-metallurgy of zinc *see* Luckow, *Jour. Chem. Soc.*, Vol. XLII., 1882, p. 431 ; and the separation of zinc and silver, by W. G. Blagden, *see* Watts's "Dictionary of Chemistry," Vol. VI., p. 1,026.

Separation of Magnesium.—Mg. Electro-chemical equivalent $\frac{24\cdot3}{2} = 12\cdot15$. A dyad cation. Fused magnesic chloride is not reduced to metal by contact with aluminium (Flavitzky, Watts's "Dictionary of Chemistry," Vol. VIII., Part L, p. 64). An amalgam of potassium or sodium decomposes a solution of magnesic sulphate by simple contact, and produces magnesium amalgam. Electrolysis of that liquid into a cathode of mercury also produces it (Klauer, Watts's "Dictionary of Chemistry," Vol. III., p. 888 ; *Jour. Chem. Soc.*, 1876, Part L, p. 684). I melted to a perfect liquid, at nearly a white heat, a mixture of six grains of magnesic fluoride and four of calcic fluoride, and then added two grains of crystals of silicon. The crystals did not dissolve, and there appeared no signs of magnesium having been separated. The electrolytic decomposition of the double chloride of magnesium and sodium in a fused state by contact with sodium is the ordinary process of obtaining the metal.

Magnesium is a very highly electro-positive metal ; it deposits in the metallic state nearly all the base and noble metals from solutions of their salts by simple contact. According to Roussin, it deposits bismuth, platinum, gold, silver, mercury, copper, lead, thallium, tin, and cadmium (*Chem. News*, Vol. XIV., p. 27). In addition to these, according to Phipson, it deposits nickel, cobalt, and zinc, and even iron and manganese, from solutions of ferrous and manganous salts; but it does not deposit aluminium from its solutions (Watts's "Dictionary of Chemistry," Vol. V., p. 795).

A sub-oxide of magnesium appears to be formed when sodic or ammonic chloride is electrolysed with electrodes of magnesium wire, the anode being covered with the black oxide (W. Beetz, Watts's "Dictionary of Chemistry," Vol. VI., p. 796). I have obtained this compound in a great variety of liquids by immersing magnesium in contact with platinum or palladium in them; solutions of chloride and bromide of potassium or sodium were some of the most suitable liquids (*Proceedings* Birmingham Philosophical Society, Vol. IV., Part I.).

The metal is deposited by means of a separate current. According to Bertrand, an adherent deposit of the metal may be obtained by electrolysing during a few minutes a concentrated aqueous solution of the double chloride of magnesium and ammonium by means of a very powerful current and a cathode of copper (*Chem. News*, Vol. XXXIV., p. 227 ; *Jour. Chem. Soc.*, 1877, Part L, p. 161). Bunsen obtained it by electrolysing fused chloride of magnesium at a red heat in a deep and covered porcelain crucible, which was divided by a vertical partition of porous porcelain extending from the top to half way down the vessel. The current employed was from ten zinc and carbon elements. The electrodes were of carbon and were introduced through openings in the cover, and the cathode was notched, so that the light melted metal collected in the notches, instead of rising to the surface and then burning. Matthiessen states that the metal may be much more easily obtained by this method if the salt employed consists of a mixture of four molecules of magnesic chloride, three of chloride of potassium, and a little chloride of ammonium. In this case, the liquid salt being lighter than the magnesium, the latter falls to the bottom (Watts's "Dictionary of Chemistry," Vol. II. p. 438 ; Vol. III., p. 751).

For the use of electrolysis in the metallurgy of magnesium, see F. Fischer, *Jour. Chem. Soc.*, Vol. XLIV., 1883, p. 399.

Separation of Thorium.—Atomic weight = 233·9. Sodium sets free metallic thorium from fused double chloride of thorium and potassium in an iron crucible (L. F. Nilson, *Jour. Chem. Soc.*, Vol. XLIV., 1883, p. 152).

Separation of Norwegium.—Atomic weight = 145·9 (?). According to Dr. Tellef, the sulphate solution of this metal is turned brown on the addition of zinc, and the metal is deposited in the pulverulent state (*Chem. News*, Vol. XL., p. 25).

Separation of Cerium. Ce. Atomic weight = 94·2. **Lanthanum.** La. Atomic weight = 92. And **Didymium.** Dy. Atomic weight = 96.—Crude double chloride of cerium and potassium in a fused state at a red heat is decomposed by metallic sodium, and metallic globules of impure cerium, together with shining scales of an oxychloride of cerium, are obtained (Wohler, Watts's "Dictionary of Chemistry," Vol. VI., p. 419).

When a mixture of oxide of cerium and potassic fluoride is melted in a porcelain crucible, and subjected to electrolysis, potassium and silicide of cerium in the form of a brown mass are deposited upon the cathode (Ulik, Watts's "Dictionary of Chemistry," Vol. V., p. 266 ; and Vol. VI., p. 420).

According to Bunsen, either of these metals may be separated by electrolysis with a separate current in the following manner :—Its chloride is mixed with sal-ammoniac (both as dry as possible), and the mixture heated to redness in a platinum crucible to expel all the sal-ammoniac. A porous clay vessel of the best quality is filled with the residue, then placed in a Hessian crucible, surrounded by a cylinder of sheet iron (with a long projecting strip for connection) to serve as the anode, and the space between the two vessels filled with a previously melted mixture of an equal number of equivalents of the chlorides of potassium and sodium. A thick iron wire, enclosed in a clay pipe, has a coil of very fine iron wire at its extremity to serve as the cathode, and is immersed in the fused salt in the inner vessel. The fusion is effected by preference in a fire of glowing charcoal, to prevent as far as possible the presence of aqueous vapour, and a strong current is employed (*Electrical News*, Vol. L, p. 184 ; *see* also Hillebrand and Norton, Watts's "Dictionary of Chemistry," Vol. VIII., pp. 420-421).

T. Schuchardt states that he has succeeded in obtaining by electrolysis metallic cerium in globules weighing from four to five grammes ; and that he has also by the same process, with the aid of a current from six Bunsen cells, obtained metallic didymium in globules the size of a pea (*Chem. News*, Vol. XL., p. 35). Hillebrand and Norton also state that they have obtained each of these metals by means of electrolysis (*Jour. Chem Soc.*, 1876, Part II., p. 276).

According to C. Erk, the electrolysis of a neutral solution of cerous nitrate by means of a current from three Bunsen cells yielded at the cathode a brownish-yellow mass and a quantity of ammonia sufficient to precipitate the whole of the cerium. A concentrated one of cerous chloride gave free chlorine at the anode and a deposit of ceroso-ceric hydrate at the cathode. The same salt in a state of fusion, with an anode of gas-carbon, gave small quantities of metallic cerium, and reddish white laminæ of cerium oxychloride at the cathode; and at the anode hydrochloric acid was evolved, and a large quantity of ceroso-ceric oxide formed. Strong solutions of cerous sulphate became yellow at the anode from formation of ceroso-ceric sulphate, and at the cathode yielded a little metallic cerium and a waxy deposit of ceroso-ceric sulphate, which subsequently became crystalline. An aqueous solution of cerous acetate yielded a basic acetate (Watts's "Dictionary of Chemistry," Vol. VII., p. 274).

Separation of Gallium.—Ga. Atomic weight = 69·86. Gallium is allied to aluminium and also to mercury. Cadmium separates gallium in the metallic state from a boiling solution of its chloride by prolonged immersion (M. Lecoq de Boisbaudran, *Jour. Chem. Soc.*, Vol. XLII., 1882, p. 897). So long as the liquids are sensibly acid, and the evolution of hydrogen goes on actively, zinc does not precipitate either the chloride or sulphate of gallium; but when the liquids become basic, and hydrogen is evolved but slowly, either the oxide or a subsalt of gallium separates in white flakes mixed with subsalts of zinc (M. Lecoq de Boisbaudran, *Chem. News*, Vol. XXXV., p. 158).

On passing a current from five bichromate cells through an ammoniacal solution of sulphate of gallium, with platinum electrodes, metallic gallium is deposited on the cathode, and a white film is formed upon the anode. In four and a-half hours the metallic deposit weighed ·0016 gramme. With ten cells, in five hours it weighed ·0034 gramme. The metal was adhesive, not easily burnished by friction, but better by pressure (Lecoq de Boisbaudran, *Jour. Chem. Soc.*, 1876, Part L, p. 521 ; *Chem. News*, Vol. XXXV., p. 150).

According to Schicht, by electrolysis, gallium, like zinc, is deposited completely and in a pure state upon the cathode (*Chem. News*, Vol. XLI., p. 280). By the electrolysis of a solution of oxide of gallium in one of caustic potash, by means of a current from five or six Bunsen's cells, and platinum electrodes, metallic gallium is deposited as liquid globules (M. Lecoq de Boisbaudran, *Chem. News*, Vol. XXXV., pp. 150 and 168).

Separation of Aluminium.—Al. Electro-chemical equivalent = $\dfrac{27\cdot5}{3}$ = 9·16. A triad cation. Magnesium by simple immersion in solutions of aluminium salts produces aluminic hydrate (S. Kern, *Chem. News*, Vol. XXXIII., p. 236). Magnesium does not deposit aluminium as metal from its solutions (Roussin, *Chem. News*, Vol. XIV., p. 27). An amalgam of aluminium is formed by contact of the metal with sodium amalgam and water (Watts's "Dictionary of Chemistry," Vol. III., p. 886).

Various persons have stated that aluminium cannot be deposited by the aid of a separate electric current. "The electrolytic reduction of aluminium may be performed either in the wet or in the dry way. The reduction from the fused chloride of aluminium and sodium was first effected by Bunsen in 1854. The salt is introduced in a fused state into a red-hot porcelain crucible, divided into two parts by a porous earthenware diaphragm, and the extremities of the poles of a Bunsen battery of ten elements are introduced into the two halves of the fused mass. The metal is then reduced at the cathode, and if the temperature is sufficiently high the metal is melted into globules" (Watts's "Dictionary of Chemistry," Vol. I., p. 152).

M. H. St. Claire Deville says :—" It appeared to me impossible to obtain aluminium by the battery in aqueous liquids. I should believe this to be an impossibility if the brilliant experiments of M. Bunsen on the production of barium did not shake my conviction. Still, I may say that all processes of this description which have recently been published for the preparation of aluminium have failed to give me good results. It is of the double chloride of aluminium and sodium, of which I have already spoken, that this decomposition is effected. The bath is composed of two parts by weight of chloride of aluminium, with the addition of one part of dry and pulverised common salt; the whole is mixed in a porcelain crucible, and heated. The combination is effected with disengagement of heat, and a liquid is obtained which is very fluid at 392° F., and fixed at that temperature. It is introduced into a vessel of glazed porcelain, which is to be kept at a temperature of about 329°F. The cathode is a plate of platinum on which the aluminium (mixed with common salt) is deposited in the form of a greyish crust. The anode is formed of a cylinder of charcoal, placed in a perfectly dry porous vessel, containing melted chloride of aluminium and sodium. (The densest charcoal rapidly disintegrates in the bath, and becomes pulverulent; hence the necessity of the porous vessel.) The chlorine is thus removed with a little chloride of aluminium proceeding from the decomposition of the double salt. This chloride would volatilise and be entirely lost if some common salt were not in the porous vessel. The double chloride becomes fixed, and the vapours cease. A small number of voltaic elements (two are all that are absolutely necessary) will suffice for the decomposition of the double chloride, which presents but little resistance to the electricity. The platinum plate is removed when it is sufficiently charged with the metallic deposit. It is suffered to cool, the saline mass is rapidly broken off, and the plate replaced" (*The Chemist*, New Series, No. XIII., October, 1854, p. 12). By the electrolysis of fused sodic aluminic chloride the aluminium deposited contains silicium derived from the charcoal electrodes (Deville, Watts's "Dictionary of Chemistry," Vol. I, p. 152, and Vol. V., p. 267).

According to A. Bertrand, by means of a separate current aluminium is deposited on a copper plate in granules from aluminium ammonium chloride, whilst chlorine is evolved at the anode (*Jour. Chem. Soc.*, 1877, Part I, p. 161; *Chem. News*, Vol. XXXIV., p. 227). M. Corbelli deposits the metal by electrolysing a mixture of rock alum, or sulphate of aluminium, and the chlorides of calcium or of sodium, the anode being formed of iron wire coated with an insulating material, and dipping into mercury placed at the bottom of the solution, and the cathode of zinc immersed in the solution. Aluminium is then deposited upon the zinc, and the chlorine which is eliminated at the anode unites with the mercury and forms m (Watts's "Dictionary of Chemistry," Vol. I., p. 152)calo el

Thomas and Tilley state that they deposit aluminium from a solution composed of freshly precipitated alumina dissolved in boiling water containing cyanide of potassium; also from a solution of calcined alum in aqueous cyanide of potassium, and from several other liquids. They also state that they have deposited alloys of aluminium and silver; aluminium, silver, and copper; aluminium and tin; aluminium, silver, and tin; aluminium and copper; aluminium and nickel; aluminium and iron, &c. J. B. Thompson says that he has for more than two years been depositing aluminium on iron, steel, and other metals, at a temperature of about 500° F., and also depositing aluminium bronze of various tints from the palest yellow to the richest gold colour (*Chem. News*, Vol. XXIV., p. 194). Jeancon deposits the metal from an aqueous solution of a double salt of aluminium and potassium of specific gravity 1·161, at a temperature of 140° F., by means of a current from three Bunsen cells (*Telegraphic Journal*, Vol. I, p. 308). T. Ball also deposits it from the double chloride of aluminium and potassium (*Chem. News*, Vol. V., p. 153).

I electrolysed a strong solution of aqueous fluoride of aluminium, containing free hydrofluoric acid, with large sheet platinum electrodes and a strong current. Gas was evolved freely from the anode, and the liquid became heated.

Aluminium used as an anode in dilute sulphuric acid largely stops the current, probably by becoming coated with a layer of insulating oxide; but if employed only as a cathode it is not thus effected (*Chem. News*, Vol. XXXI., p. 99; *Telegraphic Journal*, Vol. III., p. 59).

It may be superficially coated with mercury by being made the cathode in contact with mercury in acidulated water

(Cailletet, *Comptes Rendus*, XLIV., p. 1,250 ; also Watts's "Dictionary of Chemistry," Vol. VII., p. 54).

Aluminium, like magnesium, has great power in reducing metallic solutions and depositing their metals by simple immersion process ; it reduces those of silver, mercury, copper, lead, thallium, and zinc (*see* A. Cossa, Watts's "Dictionary of Chemistry," Vol. VII., p. 54).

For the use of electrolysis in the metallurgy of aluminium, *see* F. Fischer, *Jour. Chem. Soc.*, Vol. XLIV., 1883, p. 399. For the electrolytic analysis of compounds of aluminium, *see* V. Francken, *Chem. News*, Vol. XLVI., p. 106; also *Jour. Chem. Soc.*, Vol. XLII., 1882, p. 132 ; and A. Claessen, *ibid.*, p. 896.

Separation of Glucinum.—Gl. Atomic weight = 9·3. A cation. Nilson and Petterson were unable to separate this metal by the separate current method (*Chem. News*, Vol. XXXVII., p. 195). Becquerel deposited the pure metal from a concentrated solution of its chloride by means of a current from twenty voltaic cells. It was in the form of brilliant, steel grey crystalline laminæ (Gmelin, "Handbook of Chemistry," Vol. III., p. 293).

For the electrolytic analysis of its compounds, *see* A. Claessen, *Jour. Chem. Soc.*, Vol. XLII., 1882, p. 896.

Separation of Calcium.—Ca. Atomic weight = 40. According to Klauer, calcium amalgam may be formed either by simple immersion of sodium amalgam in solutions of calcium salts, or by passing a strong electric current from those liquids into mercury. Herschel observed that during the electrolysis of a solution of calcic chloride by means of a separate current, the cathode evolved gas and became coated with caustic lime.

This metal was first separated by an electric current during the year 1808 by Sir H. Davy, who obtained it as an amalgam by employing a cathode of mercury. Fremy subsequently electrolysed pure calcic fluoride in a fused state in a platinum crucible. Brisk effervescence occurred in the mass, a gas was set free at the anode, metallic calcium was deposited upon the cathode and became converted into lime by the oxygen of the air. It was difficult to make the observations, and the crucible was soon alloyed and perforated by the action (*The Chemist*, New Series, Vol. II., p. 548).

Matthiessen electrolysed a fused mixture of two molecules of calcic and one of strontic chloride with a small amount of sal-ammoniac in a porcelain crucible. The anode was of gas carbon, and the cathode was formed by winding a thin iron wire round a thicker one and dipping its end only just into the liquid. The calcium was set free as metallic globules upon the thin wire. He states that the metal deposited upon the cathode by a separate current in a fused mixture of chloride of calcium and the chlorides of potassium or sodium is not calcium (Watts's " Dictionary of Chemistry," Vol. L, p. 715).

Bunsen deposited calcium in a similar manner to that employed for manganese (*see* paragraph on " Separation of Manganese "), except that he used a greater density of current. He acidulated a concentrated and boiling hot solution of the chloride with hydrochloric acid, poured the boiling liquid into the porous cell, and employed as a cathode an amalgamated platinum wire. The calcium was deposited as a grey layer upon the amalgamated surface. The process is difficult, because the calcium quickly oxidizes to a layer of lime, which covers the cathode and stops the current. The deposit must be frequently removed, and the wire freshly amalgamated each time before re-immersion ; and even then but a small amount of the metal is obtained (*The Chemist*, New Series, Vol. L, Part II., p. 686, August, 1854).

Separation of Strontium. — Sr. Atomic weight = 87·5.

A cation. Solutions of salts of strontium are slowly decomposed by simple immersion of metallic magnesium ; after two days they yield a white deposit of strontium hydrate (S. Kern, *Chem. News*, Vol. XXXIII., p. 112 ; *Jour. Chem. Soc.*, 1876, Part L, p. 684). Sodium amalgam decomposed a saturated solution of chloride of strontium with formation of strontium amalgam (Watts's " Dictionary of Chemistry," Vol. III., p. 886 ; *see* also Vol. VIII., Part IL, p. 1,829). Silicon does not separate strontium from heated fluoride of strontium. I heated to redness a mixture, of the two substances, but no chemical change occurred. Caron deposited the metal by fusing its chloride with an alloy of sodium with tin or lead ; the reduction was not effected by sodium alone (Watts's " Dictionary of Chemistry, Vol. V., p. 436). Strontium is electro-positive to magnesium, but not to potassium or sodium, in water.

Sir H. Davy was the first to deposit this metal by means of a separate current. He formed into a cup a pasty mass of strontium carbonate with water, placed the cup upon a platinum dish, and filled the cup with mercury as the cathode. By passing a current from 300 voltaic cells from the platinum to the mercury the strontium was deposited upon and absorbed by the mercury. Hare obtained the metal in a similar manner (Watts's " Dictionary of Chemistry," Vol. V., p. 436).

Bunsen obtained strontium in a precisely similar way to that of obtaining manganese (see *ante*), using a salt of strontium instead of one of that metal (Watts's " Dictionary of Chemistry," Vol. IL, p. 437). Matthiessen obtained it from the fused chloride in the following manner :—A small porous cell was placed in a porcelain crucible, and both vessels nearly filled with anhydrous chloride of strontium, the level of that in the porous cell being the highest. The salt was melted so that a crust appeared on its surface. The cathode consisted of a thick iron wire, enclosed in the stem of a tobacco pipe, so that only 1-20th of an inch of it projected at the lower end, round which a very thin iron wire was coiled. The anode was a cylinder of sheet iron placed in the outer space. The cathode was immersed in the inner vessel, and the current passed; the metal collected upon it beneath the crust (Watts's " Dictionary of Chemistry," Vol. II., p. 438).

Separation of Barium.—Ba. Atomic weight = 137. A cation. Sodium amalgam separates this metal from a saturated solution of barium chloride at 93° C., and forms an amalgam (Crookes, *Chem. News*, Vol. VI., p. 194 ; Watts's "Dictionary of Chemistry," Vol. VI., pp. 252, 253).

Barium amalgam may be prepared electrolytically either by depositing barium into mercury, or by contact of sodium amalgam with solutions of chloride of barium. It is a soft, pasty substance, somewhat gritty (Cailletet, Watts's " Dictionary of Chemistry," Vol. III., p. 886).

Sir H. Davy was the first to deposit barium by means of a separate current. He employed a wet mass of barium hydrate, carbonate, chloride, or nitrate, a cathode of mercury, and a powerful current from 500 voltaic cells, and obtained the metal as an amalgam with the mercury. Hare prepared it in a similar manner from the chilled and moistened chloride by means of a current from 100 cells (Watts's "Dictionary of Chemistry," Vol. L, p. 500). Bunsen electrolysed a boiling hot concentrated and acidulated solution of chloride of barium in a similar way to the one he employed in obtaining manganese, chromium, and calcium. It was more easily obtained than calcium (see *ante*; also *The Chemist*, New Series, Vol. I., p. 686 ; Watts's "Dictionary of Chemistry," Vol. L, p. 500). Matthiessen obtained barium from its fused chloride in a similar manner to that in which he obtained strontium (see *ante*; also Watts's " Dictionary of Chemistry," Vol. II., p. 438).

A solution of barium nitrate, electrolysed with platinum electrodes, yielded nitric acid at the anode and caustic baryta at the cathode (Sir H. Davy).

Separation of Lithium.—Li. Electro-chemical equivalent = 7. A monad cation. I observed that lithium was not separated by adding crystals of silicon to a fused mixture of the fluorides of lithium and sodium, nor were the crystals corroded or altered in weight. I also fused some fluoride of lithium in an open platinum crucible within a partially covered clay muffle, and electrolysed it by means of a current from six Smee elements, and two flat platinum wire helices as electrodes, during thirty minutes. The conduction was free, and much gas was evolved from the anode only all the time. The anode was not corroded. A small amount of lithium was :

deposited upon the platinum cathode, and alloyed with it. By electrolysing a larger mass of the salt, with a current from six Grove cells and a thick platinum wire cathode enclosed within, but insulated from a platinum tube, to exclude the air from contact with the deposited lithium, the action was copious ; with a gold anode the gold was corroded freely, and particles of it in large quantity floated in the liquid and united the electrodes. The cathode swelled greatly, and its lower end bent itself towards the anode, became quite grey in colour, and split in the direction of its length.

Bunsen was the first person who electro-deposited this metal (Watts's "Dictionary of Chemistry," Vol. III., p. 727). By electrolysing fused chloride of lithium with a current from four or six Bunsen cells, an anode of gas coke, and a cathode of iron wire, he deposited silver white metal upon the wire, (Watts's "Dictionary of Chemistry," Vol. II., p. 437). Schnitzler also electrolysed a mixture of the fused chlorides of lithium and ammonium by a current from twelve Bunsen cells, and a cathode of iron wire, and obtained a metallic lithium (*Jour. Chem. Soc.*, Vol. XXII., p. 961).

Separation of Sodium.—Na. Electro-chemical equivalent $= 23$. A monad cation. In a solution of sodic chloride, magnesium evolves hydrogen slowly, sodium hydroxide being formed, rendering the solution alkaline (S. Kern, *Jour. Chem. Soc.*, 1876, Part L, p. 684). Beetz observed that under these conditions a black suboxide of magnesium is formed. Carbon, also iron, reduces the melted hydrate or carbonate of sodium at a high temperature, and sets free the metal.

Sir H. Davy first electro-deposited sodium in the year 1807 by moistening its hydrate with water in a platinum capsule which acted as the anode, dipping a platinum wire cathode in the salt, and using a current from a battery composed of 100 to 200 cells. He also deposited it more easily into mercury in a similar way to that already described under magnesium, and thus obtained an amalgam of the two metals.

In the electrolysis of melted sodic hydrate an anode of either platinum, silver, or copper dissolves in the liquid, and the respective metals are deposited upon the cathode (A. Bréster, *Chem. News*, Vol. XVII., p. 145).

Electrolysis of Sodic Fluoride.—Na.F. Molecular weight $= 42$. I have noticed that crystals of silicon thrown into melted fluoride of sodium evolved bubbles of vapour, which exploded and burned with a yellow flame on arriving at the surface of the liquid. In a second trial, 7 grains of the dry fluoride in powder mixed with one grain of the crystals were heated to redness; the crystals lost ·15 grain in weight. I electrolysed a saturated aqueous solution of sodic fluoride by a current from six Grove cells with platinum electrodes; gas was evolved from the anode, and emitted a powerful odour of ozone.

Electrolysis of Sodic Chloride.—Na.Cl. Molecular weight = 58·5. Hisinger and Berzelius electrolysed a solution of common salt with silver electrodes. Gas was evolved at the cathode, and after a time at the anode also. The anode became covered with argentic chloride, the liquid near it contained dissolved chlorine, and the solution near the cathode contained free soda. With lead electrodes the negative wire evolved gas, and received a deposit of crystals of lead, and the anode became coated with plumbic chloride. By electrolysing a solution of common salt, Higgins and Draper observed that chlorine was set free at the anode, and hydrogen gas and soda at the cathode. But if the cathode consisted of mercury sodium amalgam was ·produced. According to Matthiessen, a fused mixture of the chlorides of calcium and of sodium yields a deposit of the latter metal, when electrolysed in a certain manner (Watts's "Dictionary of Chemistry," Vol. I., p. 715).

Electrolysis of Sodium Carbonates.—Na_2CO_3 and NaHCO_3. According to Favre and Roche, by electrolysis, neutral sodium carbonate splits up into $CNa.O_3$ and Na, the sodium being oxidized by the water with separation of hydrogen. The acid carbonate is resolved into Na and CHO_3, the sodium being then oxidized and hydrogen evolved; the $2CHO_3$ is then resolved into $2CO_2 + H_2O + O$. According to Burckhard, sodic carbonate in a state of fusion is a good conductor, and is decomposed by electrolysis into carbonic acid at the anode, and soda together with a little carbon at the cathode (Chem. News, Vol. XXL, p. 238).

Electrolysis of Biborate of Sodium.—Fused borax yields oxygen gas at the anode and boron at the cathode. The boron is separated by indirect action ; the current resolves the soda into oxygen and sodium, and the latter separates boron from the boracic acid (Faraday, Gmelin's " Handbook of Chemistry," Vol. L, p. 460). Burckhard states that fused borax conducts, suffers electrolysis, and a series of compounds are formed or volatilised ; but the chief result is that the salt is decomposed into soda and boron at the cathode and oxygen at the anode (Chem. News, Vol. XXL, p. 238).

Electrolysis of Sodic Sulphate. — Na_2SO_4. Molecular weight = 140. By the electrolysis of this salt in a fused state with platinum electrodes, sodium is deposited and combines with the cathode (Brester, Chem. News, Vol. XVIII., p. 154). From the results obtained by electrolysing sulphide of sodium. Buff concluded that all the sulphur travelled to the anode and the sodium towards the cathode (*Chem. News,* Vol. XV., p. 279).

Electrolysis of Diphosphate of Sodium.—Na_2HPO_4. A solution of this substance is decomposed by a separate current into phosphoric acid at the anode and soda at the cathode. According to Faraday, acid phosphate of sodium in a state of fusion yields hydrogen at the cathode (Gmelin, "Handbook of Chemistry," Vol. L, p. 460). According to Burckhard, fused pyrophosphate of sodium, electrolysed with platinum electrodes, yields phosphide of platinum; but the chief result is that the salt splits up into oxygen, phosphorus, and soda (*Chem. News,* Vol. XXL, p. 238).

For the reducing action of sodium amalgam on solutions of silver, mercury, iron, and chromium, *see* the sections relating to those metals; also Watts's "Dictionary of Chemistry," Vol. VI., p. 816. For Jablochkoff's process of making sodium by electrolysis, see *Scientific American*, Sept. 22, 1883, p. 643.

Separation of Potassium.—K. Electro-chemical equivalent = 39·1. A monad cation. Magnesium, by simple immersion in a solution of potassic dichromate, forms potassic hydroxide (S. Kern, *Jour. Chem. Soc.*, 1876, Part II., p. 479). Zinc amalgam immersed in a solution of caustic potash liberates pure hydrogen (Watts's "Dictionary of Chemistry," Vol. III., p. 891). According to W. Skey, an aqueous solution of potassic chloride becomes alkaline by contact either with zinc or with silver, in the first case, probably by decomposition of water and formation of ammonia, aided by formation of zinc oxide, and in the second by oxidation of the silver by free oxygen, and the subsequent decomposition of that oxide with formation of silver chloride and caustic potash (*Jour. Chem. Soc.*, 1876, Part II., p. 266). Both carbon and iron separate potassium from melted potash at a white heat, and the process for obtaining potassium is based upon this fact. Brester states that even silver will dissolve in large quantities in melted potassic hydrate (*Chem. News*, Vol. XVIII., p. 145), and I have observed that when this hydrate is melted in a pure silver crucible the vessel loses in weight.

Electrolysis of Potassic Hydrate. — KHO. Molecular weight = 56·1. Potassium was first separated by electrolysis in the year 1807 by Sir H. Davy. He moistened a piece of potassic hydrate with water, placed it in a platinum capsule, which acted as a cathode, and touched the hydrate with the platinum wire anode of a battery of from 100 to 200 of Wollaston's cells. The potash liquefied, and globules of the metal separated at the cathode. Since that time it has been found that even a feeble voltaic current will liberate potassium from aqueous solutions of some of its salts, and if the deposited metal is protected from oxidation by being deposited into a cathode consisting of a large bulk of mercury with but a small portion of its surface exposed to the liquid, the potassium can be obtained in the form of its amalgam.

When mercury is placed in a cup connected with the cathode of a voltaic battery of at least 20 pairs, and covered with a strong solution of caustic potash, in which a piece of that substance is immersed, and into which the anode dips, the mercury takes up potassium (Berzelius, Watts's "Dictionary of Chemistry," Vol. III., p. 889).

According to Janeczek, when melted potash is electrolysed oxygen is evolved at the anode and potassium at the cathode, but no hydrogen; but if the experiment is made in a closed apparatus, after some time it is found that water is evolved with the oxygen, and some hydrogen is also set free; the latter probably results from the action of the potassium upon the melted alkali (*Jour. Chem. Soc.*, 1876, Part L, p. 182; also Watts's "Dictionary of Chemistry," Vol. VIII., Part I., p. 709).

Brester states that in the electrolysis of melted caustic potash an anode of either platinum, silver, or copper dissolves in the liquid, and the respective metals are deposited upon the cathode (*Chem. News*, Vol. XVIII., p. 145).

Electrolysis of Potassic Nitrate.—KNO_3. Molecular weight $= 101\cdot1$. Some information has already been given of the effects of an electric current upon this compound (*see* "Electrolysis of Oxides of Nitrogen"). According to Faraday, an aqueous solution of it conducts electricity very easily, and yields hydrogen at the cathode.

Electrolysis of Potassic Fluoride. — KF. Molecular weight $= 58\cdot1$. Fremy electrolysed fused potassic fluoride, and obtained a gas which rapidly attacked platinum, decomposed water, with formation of hydrofluoric acid, and displaced iodine from metallic iodides (Watts's "Dictionary of Chemistry," Vol. II., p. 673). The following are experiments of mine made with this substance :—I fused 130 grains of the pure salt in a platinum crucible within a partially covered clay muffle inserted in the hole in the top of a small gas furnace, and electrolysed it during two and a-half hours, by means of a current from six Smee cells, and two flat helices of platinum wire as electrodes. There was free conduction and much gas (of an odour like that of hydrofluoric acid) evolved from the anode, but none from the cathode, and no signs of any deposit. The anode was not corroded, nor altered in weight. I also electrolysed some of the same salt in a state of fusion by means of a current from six Grove cells with a thick platinum wire as the anode and the platinum vessel as the cathode. Great heat was evolved, and violent electrolytic action occurred; nearly white hot metallic globules also accumulated and exploded repeatedly. The end of the anode fused, and particles of platinum ramified from it in white hot threads, and a short electric arc (about 1-10th of an inch in length) was produced.

I also perfected and used a somewhat elaborate platinum apparatus, by means of which the gas from the anode was prevented from coming in contact with the cathode, and might be collected, the electrodes being enclosed within (but isolated from) two wide platinum tubes. One thousand grains of the perfectly pure salt were electrolysed in this apparatus by means of a current from six Grove cells. The anode, which was a solid rod of platinum, was rapidly corroded, and was thus cut off at the level of the liquid and stopped the current; the corroded surface was very bright, as if fused potassium was deposited upon the cathode. Much spongy platinum was diffused in the melted salt, and the apparatus was a little corroded at the surface of the liquid. No gas was evolved at the anode. The deposited potassium did not alloy with the stout rod of platinum used as the cathode. 55·35 grains of grey metallic platinum were found in the saline mass; a salt of platinum appeared to have been formed at the anode, then dissolved or diffused throughout the liquid and decomposed by the heat, and thus the liberated fluorine did not escape at the anode, but was evolved in the mass of the liquid generally, and came into contact with the liberated potassium.

Having ascertained the electrical relations of palladium, gold, platinum, and iridium in the fluoride, palladium being the most positive and iridium the most negative, I repeated the experiments with an anode of iridium and a current from three Grove cells. Copious clouds descended at once from the anode, and made the liquid opaque; there was also a violent action at the anode. The anode became black, and a little gas was evolved from it, accompanied by an acid odour like that of a mixture of sulphurous anhydride and hydrofluoric acid. Potassium was freely liberated at the cathode, and produced occasional explosions. With a current from six cells the anode dissolved rapidly, and soon lost thirty-eight grains. I then put a pure gold anode, and employed two cells. Gas, of a feebly acid odour, was freely evolved at the anode; and with a current from six cells was very copious, and smelt much like sulphurous anhydride. The gold dissolved much less rapidly than the iridium. With a palladium anode and a current from six cells the anode rapidly dissolved, potassium was deposited and exploded frequently, and an odour like that of hydrofluoric acid was strong, much gas being liberated; 33·3 grains of free metal were found in the saline mass. The platinum cathode was not corroded.

In the experiments the platinum anode dissolved, as if melted; the iridium one was black, the palladium one was oxidised of various colours. The platinum vessel was cut into at the level of the surface of the liquid, evidently not by the fused fluoride of potassium, but by some substance set free at the anode by electrolysis. In another instance I electrolysed the pure fused fluoride with a large platinum anode, small platinum cathode, and a current from three Grove cells during half an hour. Much gas, having an odour of ozone and hydrofluoric acid, was evolved from the anode, and the latter dissolved rapidly and lost 37·5 grains in weight. The gas reddened test paper. The platinum containing vessel was corroded at the line of surface of the liquid, and lost about eleven grains. About fifty-one grains of free metallic platinum in loose powder were found in the saline residue. Each of these experiments shows that a very corrosive substance was liberated at the anode.

I electrolysed the fused salt with a gas carbon anode and a platinum wire flat helix as a cathode with a current from six Smee cells. Free conduction occurred, and much gas was set free from the anode only. The part of the anode in the liquid was not visibly corroded.

I also electrolysed about 8oz. of pure double fluoride of hydrogen and potassium (KF HF) in a fused state during half an hour, at about 300° F., with a current from ten Smee cells, and electrodes of stout sheet platinum. There was copious conduction, and abundance of hydrogen evolved at the cathode, but no gas from the anode, which was rapidly corroded away, with a rough surface, and lost 9·37 grains. The salt became less fusible by loss of hydrofluoric acid, which escaped freely all the time. The saline residue contained a small amount of dissolved platinum salt, and nearly 9 grains of free metallic platinum. In a second experiment, lasting half an hour, the salt was kept only just fused, and a small gold anode was employed. The conduction was free, and much gas was evolved from the cathode, and a film of bright yellow gold spread over the surface of the salt, and connected the electrodes, unless the liquid was continually stirred. The anode rapidly dissolved (more quickly than that of platinum), and the salt of gold at once decomposed, and set free finely-divided gold as a dull, red-brown powder at the anode. No gas appeared at the anode at any time; that from the cathode detonated on applying a light. There was loose red-brown powder of gold, weighing 1·4 grain, upon the cathode, but of adherent gold only ·05 grain. The anode was corroded, and lost 6·80 grains. The saline residue contained no dissolved gold; but 5·85 grains of red-brown powder, containing 5·30 grains of gold.

In a third similar experiment, by using a large sheet

platinum anode and a small platinum cathode, and a current from ten Smee cells during two hours, the phenomena were the same as in previous experiments. The anode lost 28 grains ; much loose platinum collected on the cathode, which was neither corroded nor alloyed. The saline residue contained a trace of dissolved platinum salt, and nearly all the corroded platinum in a metallic state. In a fourth experiment I continued the action during three and a-half hours ; the results were as before. The loss of the anode was 35·73 grains. The saline residue contained a small quantity of dissolved double fluoride of platinum and potassium, which, after being well washed, was dried and heated to redness ; it then shot about as if gas was evolved from it. In a fifth similar experiment, lasting four and a-half hours, at the lowest possible fusion temperature, more of the brown platinum salt formed at the anode and dissolved in the liquid. The anode lost 64·81 grains. In a last experiment I electrolysed a gently fused mixture of 900 grains of the pure double salt and 100 grains of pure argentic fluoride, with a large anode of platinum and a large cathode of silver. Conduction was complete with ten Smee cells. No gas was evolved at either electrode. The surface of the anode disintegrated rapidly and lost 49·84 grains in four and a-half hours' action. The separated platinum dissolved only to a small extent in the liquid, and subsided in admixture with the silver to the bottom of the vessel as a fine, black powder, weighing 73·93 grains, which lost less than two per cent. when heated to redness. Some grey silver powder was deposited upon the cathode. In all these experiments with the acid fluoride, saline films continually formed upon the surface of the liquid. They came from the cathode and were more abundant the deeper the cathode was immersed.

I electrolysed a nearly saturated aqueous solution of pure fluoride of potassium by means of a current from six Grove cells with large platinum electrodes. Conduction was copious, and the liquid acquired a nearly boiling temperature. Much gas, having an odour like that of a mixture of ozone and chlorine, was evolved at the anode. A saturated solution of the same salt, electrolysed by a current from ten large Smee cells with large platinum electrodes, evolved gas at each electrode. That from the anode smelt powerfully of ozone, and reinflamed a red hot splint. Several other experiments with variations in the size of the electrodes were made, and with addition of hydrofluoric acid, but the results were similar.

I saturated some pure dilute hydrofluoric acid of 40 per cent. at 60° F. with pure double fluoride of hydrogen and potassium, and electrolysed the solution by a current from ten Smee cells, a gold anode, and a platinum cathode, during 5½ hours. Gas was evolved freely from both electrodes, and a strong odour of ozone was observed. The anode lost 1·73 grain, and the cathode acquired first a gilded appearance and then a black coating, and the liquid became black with finely divided matter.

Electrolysis of Potassic Chloride.—KCl. Molecular weight = 74·6. Matthiessen separated the metal from this salt by means of a current from six Bunsen's elements with carbon electrodes. He melted a dry mixture of one molecular weight of potassic chloride and one of calcic chloride, arranging the distribution of the heat so that a little of the mixture remained unfused around the upper part of the cathode. Chlorine escaped at the anode, and pure potassium accumulated around the anode beneath the crust. The crucible was then cooled, and its contents removed under rock oil (Watts's "Dictionary of Chemistry," Vol. IV., p. 692.)

Electrolysis of Potassic Chlorate.—$KClO_3$. Molecular weight = 122·6. According to Brester, the electrolysis of melted potassic chlorate, with a platinum anode, yielded potassium, which united with a cathode of copper or platinum. Chlorine and oxygen, with an odour of phosphorus, were set free at the anode, and formed thick white vapours by contact with water (*Chem. News*, Vol. XVIII., p. 145).

Electrolysis of Potassic Iodide.—KI. Molecular weight = 166·1. According to H. St. Claire Deville, silver renders a solution of potassic iodide alkaline by simple immersion in it; it also liberates potassium by a similar reaction when immersed in the fused salt (*The Chemist*, New Series, Vol. IV., p. 329). I have noticed that mercury, by prolonged contact with a perfectly neutral solution of the salt, renders it faintly alkaline. Faraday found that by passing an electric current through melted iodide of potassium iodine was set free at the anode and potassium at the cathode.

Electrolysis of Potassic Carbonate.—K_2CO_3. Molecular weight = 138·2. By the electrolysis of solutions of hydropotassic carbonate, carbonic anhydride is very incompletely evolved at the anode (C. Luckow, *Jour. Chem. Soc.*, Vol. XXXVIII., 1880, p. 283). According to Faire and Roche, in the electrolysis of solutions of alkaline carbonates or bicarbonites, the molecule splits up in such a way that an atom of potassium or sodium is set free at the cathode, and liberates hydrogen (*Chem. News*, Vol. XXX., p. 63; *Jour. Chem. Soc.*, Vol. XII., p. 861).

Electrolysis of Potassic Cyanide.—KCy. Molecular weight = 65·1. According to H. St Claire Deville, even platinum, in a state of fine powder, when immersed in a boiling hot solution of this salt, liberates hydrogen. By the electrolysis of a solution of potassic cyanide, Kolbe observed that potassic cyanate was formed at the anode (Watts's "Dictionary of Chemistry," Vol. II., p. 190). Faraday noticed that the aqueous solution yielded by electrolysis hydrogen and potash at the cathode, but no oxygen at the anode; that the liquid around the anode became brown, that the fused salt, and that aqueous solutions of potassic sulphocyanide and ferrocyanide behaved similarly (Gmelin's "Handbook of Chemistry," Vol. I., p. 458).

Electrolysis of Potassic Ferrocyanide.—KFCy. Molecular weight = 368. When a solution of ferrocyanide of potassium is decomposed by an electric current, ferrocyanide of potassium is formed at the anode, and hydrogen and potash appear at the cathode (Watts's "Dictionary of Chemistry," Vol. II., p. 240). The alkaline ferrocyanides yield alkali at the cathode, and hydrocyanic acid and Prussian blue at the anode, unless the anode is composed of copper, in which case a deposit is there formed of cyanide of copper (Porrett, *ibid.*, p. 222)

Electrolysis of Potassic Ferridcyanide.—KFdCy. Molecular weight = 329. Carbon charged with hydrogen easily reduces a solution of ferri to ferro cyanide of potassium (Gladstone and Tribe, *Jour. Chem. Soc.*, Vol. XXXIII., 1878, p. 309). A platinum hydrogen couple does the same readily (*ibid.*). Bottger observed a similar effect with palladium containing occluded hydrogen.

When a solution of potassic ferricyanide is electrolysed by a separate current yellow prussiate is formed upon the cathode (Watts's "Dictionary of Chemistry," Vol. II., p. 247).

Separation of Rubidium.—Rb. Atomic weight = 85·48. A monad cation. Like sodium and potassium, this metal is separated from its fused carbonate at a white heat by simple contact with carbon. It was first obtained by electrolysing its fused chloride with a graphite anode and a cathode of iron wire. It has been obtained by electrolysing a fused mixture of the chlorides of rubidium and calcium in their equivalent proportions at a temperature a little below redness. It is also obtained as an amalgam by electrolysing a strong neutral aqueous solution of rubidium chloride with an anode of platinum and a cathode of mercury. The metal itself is decidedly more electro-positive than potassium, and both it and its amalgam decompose water readily (Watts's "Dictionary of Chemistry," Vol. V., p. 129).

Separation of Caesium.—Cs. Atomic weight = 132·66. A monad cation. Unlike rubidium, potassium, and sodium, this metal is not liberated from its fused carbonate by contact with carbon at a white heat. An amalgam of the metal may be easily obtained by electrolysing a solution of caesium chloride with a cathode of mercury (Watts's "Dictionary of Chemistry," Vol. L, p. 1,114). M. Setterberg obtained metallic caesium by electrolysing a dry mixture of four parts of caesium cyanide and one of barium cyanide. This mixture fuses more easily than caesium cyanide alone (*Chem. News*, Vol. XLV., p. 94, and Vol. XLVI., p. 249).

Separation of Ammonium (?), H_4N_1, and Electrolysis of Ammonia.—H_3N. Weyl, in 1864, discovered that sodium swells, liquefies, and dissolves in anhydrous ammonia liquefied by pressure, and on removal of the pressure the sodium returns to the metallic state; also that potassium behaves similarly; that barium forms a deep blue liquid with a metallic lustre; and that silver, mercury, copper, and zinc likewise form unstable compounds with the liquefied gas (Watts's "Dictionary of Chemistry," Vol. V., pp. 328-329). Seeley in 1870 subsequently discovered that metallic rubidium, potassium, sodium, or lithium, by simple immersion in colourless anhydrous ammonia liquefied by pressure, dissolved in the liquid, and produced intensely blue solutions of powerfully reducing properties (*Chem. News*, Vol. XXIII., p. 169; Watts's "Dictionary of Chemistry," Vol. VI., p. 60). I have also verified these results, and have examined the action of the sodium solution upon various compounds (*Proc. Roy. Soc.*, Vol. XXL, 1873, pp. 140, 147). In all these blue solutions it is supposed that ammonium is set free and dissolved.

Bleekrode ascertained that anhydrous liquefied ammonia was a conductor and an electrolyte. He passed the current from 80 Bunsen cells through it; gas was evolved, and the liquid became intensely blue. He also passed the current from 3,240 cells of De la Rue's chloride of silver battery through the liquid by means of platinum wire electrodes. The anode became black, much gas was evolved, and the liquid became intensely blue. On stopping the current the colour disappeared. In these experiments ammonium was probably set free and dissolved, and produced the colour.

Damp iron filings exposed to the air or to nitrogen induce the formation of ammonia (Berzelius).

Electrolysis of Aqueous Ammonia.—By simple immersion of magnesium in solutions of ammonium salts, ammonia and nitrogen are set free (S. Kern, *Jour. Chem. Soc.*, 1876, Part II., p. 479). By electrolysis of ammonium salts ammonia is produced at the cathode (C. Luckow, *Jour. Chem. Soc.*, Vol. XXXVIII., 1880, p. 285). According to Hisinger and Berzelius a concentrated aqueous solution of ammonia conducts as imperfectly as pure water, but by addition of a little ammonic sulphate it is rendered easily decomposable. A gold anode becomes covered with amber coloured fulminate of the metal and dissolves, and the cathode is gilded. With a mixture of one volume of strong aqueous ammonia and three of water, oxygen is set free at the anode, and the anode becomes corroded. With a cathode of mercury, the bulky "amalgam of ammonium" is obtained (Gmelin's "Handbook of Chemistry," Vol. L, p. 458).

According to Favre, under the influence of the current, ammonic oxide is decomposed thus :—1st, $3 (NH_4)_2O = 3 (NH_4)_2 + O_3$. The three equivalents of ammonium set at liberty decompose the water, like potassium or sodium, thus : —2nd, $3 (NH_4)_2 + 3 H_2O = 3 (NH_4)_2O + 3 H_2$. The oxygen of equation No. 1, reacting upon the ammonium, gives :—3rd, $O + NH_4 = N + 2H_2O$. The first equation represents the electrolysis proper (*Jour. Chem. Soc.*, Vol. IX., p. 985).

The so-called "ammonium amalgam" with mercury was discovered in 1800 simultaneously by Seebeck in Jena, and by Berzelius and Pontin at Stockholm. It is obtained either by the contact of sodium amalgam with strong solution of certain salts of ammonia, or by the electrolysis of a concentrated solution of sal-ammoniac, or certain other ammoniacal salts (not the nitrate) with a cathode of mercury. In each case the mercury swells to a bulky mass, which on the cessation of electrolysis spontaneously decomposes into liquid mercury and a mixture of two volumes of ammonia and one of hydrogen. It solidifies below 0°C., and crystallises in cubes. It does not decompose below 29°C. if previously frozen. In the separate current process oxygen is evolved at the anode, if the salt employed is aqueous ammonia, or the carbonate, sulphate, or phosphate, but chlorine if it is sal-ammoniac, and but little gas is set free at the cathode (Watts's "Dictionary of Chemistry," Vol. L, p. 188). If the cathode consists of spongy platinum impregnated with mercury, much gas is evolved, and no amalgam is formed (Wetherill, *ibid.*, Vol. VI., p. 103). A concentrated solution of trimethylamine hydrochloride behaves with sodium amalgam just like one of sal-ammoniac (Pfeil and Lippmann, Watts's "Dictionary of Chemistry," Vol. VI., p. 104). Various investigators consider the ammoniacal amalgam to be merely a spongy mixture of mercury and hydrogen (*see* Seeley, *Chem. News*, 1871, Vol. XXIII., p. 169 ; also Pfeil and Lippmann, Watts's "Dictionary of Chemistry," Vol. VI., p. 104).

Electrolysis of Nitrate of Ammonia.—$NH_4 NO_3$. Molecular weight $= 80$. The action of a copper zinc couple on a solution of ammonium nitrate showed that both nitrate and ammonia were produced. In the cold the nitrate, even in a solution of 20 per cent., was completely reduced to ammonia in about 24 hours, without the escape of ammonia, free or combined (Gladstone and Tribe, *Jour. Chem. Soc.*, 1878, Vol. XXXIII., p. 150).

According to Divers, dry nitrate of ammonia condenses gaseous ammonia and becomes liquid. The liquid ammoniated ammonia nitrate is a good conductor and electrolyte—ammoniacal hydrogen appearing at the cathode, and nitrogen and ammonia nitrate at the anode. Anodes of silver, copper, lead, zinc, and magnesium are dissolved as ammoniated nitrates. An anode of mercury becomes coated with an almost insoluble compound. When the anode is acted upon the evolution of nitrogen does not occur (Watts's "Dictionary of Chemistry," Vol. VII., p. 860; *Chem. News*, Vol. XXVII., p. 37; *Proc. Roy. Soc.*, Vol. XXL, p. 109). Faraday electrolysed fused nitrate of ammonium. Hydrogen gas, mixed with a little nitrogen, was evolved at the cathode. The aqueous solution of the salt similarly treated yielded the same mixture at the cathode and oxygen at the anode.

Electrolysis of Ammonium Fluoride.—H_4NF. Molecular weight $= 37$. I electrolysed this salt in a state of gentle fusion by means of a current from six Grove cells, a platinum wire anode, and a platinum sheet cathode. Conduction was copious and heat was set free. Much gas appeared at the anode, but no odour of ozone.

Electrolysis of Ammonia Chloride.—NH_4CL. Molecular weight $= 53.5$. Hisinger and Berzelius by electrolysing with silver electrodes a solution of this salt observed that oxygen was evolved at the anode and hydrogen at the cathode, and the anode became coated with argentic chloride.

Electrolysis of Ammonic Carbonate.—According to Seebeck, a moistened cup of this salt filled with mercury yields by electrolysis with the mercury as the cathode the ammoniacal amalgam. E. Drechsel electrolysed a solution of ordinary commercial ammonic carbonate, by passing a continually reversed current through it by means of platinum electrodes during eight hours. On evaporating the resulting liquid a salt was obtained crystallising in fine white needles, and containing 64·69 per cent. of platinum. About 0·1 grain of platinum was dissolved in 10 hours by the ammonium carbonate. By working the commutator more slowly the temperature of the liquid rose, and by simultaneously cooling a crystalline precipitate occurred, containing 38·6 per cent. of platinum, and was also a salt of a platinum base *(Jour. Chem. Soc.*, Vol. XXXVIII., 1880, p. 300). B. Gerdes also electrolysed a solution of ammonium carbonate by continually reversed currents from four to six Grove cells and platinum electrodes. He obtained besides ammonium nitrite and nitrate, urea, a fatty substance, and a soluble salt of platinum *(Jour. Chem. Soc.*, Vol. XLIV., p. 27).

Electrolysis of Sulphate of Ammonium.—Am_2SO_4. Molecular weight $= 132$. A solution of ammonic sulphate is decomposed by the current, acid and oxygen appearing at the anode, alkali and hydrogen at the cathode (Sir H. Davy). With iron wire electrodes, hydrogen and free ammonia appear at the cathode, and at the anode oxygen is evolved, but not until after some time; persulphate of iron also appears (Hisinger and Berzelius).

Made in the USA
Monee, IL
01 October 2025

31113027R00090